Fluid Catalytic Cracking Technology and Operations

Fluid Catalytic Cracking Technology and Operations

Joseph W. Wilson

PENNWELL PUBLISHING COMPANY
TULSA, OKLAHOMA

Copyright © 1997 by
PennWell Publishing Company
1421 South Sheridan Road/P.O. Box 1260
Tulsa, Oklahoma 74101

Library of Congress Cataloging-in-Publication Data

Wilson, Joseph W.
    Fluid catalytic cracking technology and operations / Joseph W. Wilson.
       p.   cm.
    Includes index.
    ISBN 0-87814-710-1
    1. Catalytic cracking.   I. Title
TP690.4.W55   1997   97-16152
665.5'33--dc21   CIP

All rights reserved. No part of this book may be
reproduced, stored in a retrieval system, or
transcribed in any form or by any means, electronic
or mechanical including photocopying and
recording, without the prior written permission
of the publisher.

Printed in the United States of America

03 02 01 00 99 98 97

Contents

Introduction ix

Chapter 1 **Process Overview** 1
History 1
Role in Petroleum Refining 6
Process Description 9
Commercial FCC Designs 18
Future Trends 43

Chapter 2 **Fundamentals** 47
Reactions 47
Feedstock and Feedstock Characterization 52
Yields and Products 59
Kinetics 61
Product Properties 65
Catalyst 71
Heat Balance 72
Mass Balance and Test Runs 77
Hydrogen Balance 83
Pressure Balance 85

Chapter 3 **Riser/Reactor Design and Operation** 91
Feed Injection Section 91
Riser 101
Riser Termination 104
Reactor Vessel 112
Catalyst Stripper 115
Operating Considerations 120

Chapter 4 Regenerator Design   129
Coke Combustion   129
Vessel Design   131
Air Distributor   134
Catalyst Distribution   142
Cyclones   146
Air Blower   149
Carbon Burning Kinetics   150
Regenerator Heat Removal   151
Air Heaters   155
Operating Considerations   155

Chapter 5 Flue Gas Systems   163
Flue Gas Flows and Properties   163
Flue Gas Systems   164
Flue Gas Control Valves   166
Orifice Chamber   167
Third Stage Separators   168
Turboexpanders   171
Water Sprays   174
Flue Gas Coolers   174
Electrostatic Precipitators   175
SO_x Removal   180
NO_x Removal   180
Design and Operating Considerations   181

Chapter 6 Cyclones   183
FCC Cyclone Systems   183
Cyclone Design   187
Cyclone Performance   187
Operating Considerations   192

Chapter 7 Fluidization and Standpipe Flow   195
Fluidization Fundamentals   195
Fluidization in FCC Operations   200
Catalyst Densities   204
Standpipes   205
Gamma Scans and Radiotracer Studies   217
Cold Flow Models   219
Slide Valves   219

Chapter 8   **Product Recovery**   **223**
　　　　　　　Reactor Transfer Line   223
　　　　　　　Main Fractionator   225
　　　　　　　Gas Recovery Unit   236
　　　　　　　Treating   249
　　　　　　　Operating Considerations   253

Chapter 9   **Catalyst Technology, Selection, and Monitoring**   **263**
　　　　　　　Early Catalysts   263
　　　　　　　Catalyst Technology   264
　　　　　　　Catalyst Manufacture   269
　　　　　　　Catalyst Testing and Evaluation   272
　　　　　　　Catalyst Monitoring   278
　　　　　　　Catalyst Additives   289

Chapter 10   **Troubleshooting**   **293**
　　　　　　　Troubleshooting Basics   293
　　　　　　　Problem Areas   295
　　　　　　　Turnaround Inspections   301

Glossary of FCC Terminology   **307**

Index   **319**

Introduction

Catalytic cracking was first developed more than 50 years ago. Since that time, the process has experienced almost continuous development through the present day. As a result of this constant improvement "cat cracking" has become the major conversion process in petroleum refining.

This book is intended to serve a multitude of purposes. Taken as a whole, it may be used as an introduction to catalytic cracking for engineers who are unfamiliar with the process, its operation and its role in the overall refinery. To this end, the early chapters cover the basics of the process and the reactions involved. There is also a brief history of catalytic cracking from its inception to the present day.

This handbook is, however, also intended to serve as a desk reference for engineers currently assigned to a catalytic cracking unit. Thus, many of the chapters contain detailed discussions of various aspects of design, operation and trouble shooting.

I have been involved with this process throughout a professional career of more than 20 years. I have found it to be both fascinating and frustrating. While I have gained considerable expertise in the design and operation of FCC units, the process still presents me with interesting and challenging problems.

In truth, there are enough unknowns and uncertainties in our understanding of the process to fuel numerous friendly debates among the community of FCC experts. As any regular attendee at the National Petroleum Refiners Association Q&A sessions can testify, these discussions can become quite lively, and it sometimes appears that FCC experts cannot agree on even the most basic issues. Underneath all this, however, we share a dedication to both the process and those who work with it as well as a strong belief in its continued importance to the refining industry.

Hopefully, the reader of this book will acquire some of this dedication as well as an increased technical understanding of the process. The information I have provided represents my best understanding of fluid catalytic cracking. In areas of controversy, I have attempted to present all sides in a fair and objective manner. Inevitably, however, my personal beliefs will have colored my discussion of these issues. Thus, others familiar with fluid catalytic cracking will certainly find points of disagreement. I encourage all readers of this book to listen to these other opinions with an open ear and to form their own conclusions.

The completion of this book would not have been possible without the assistance of many friends and associates in the refining industry. I am also indebted to my wife and family for their considerable patience during many long evening sessions and to my employer, Caltex Petroleum Corporation, for permission to publish this work. I also wish to express my gratitude to Caltex for their permission to incorporate material from documents prepared as a part of my employment into this book.

J.W. Wilson

Process Overview

History

The first commercial catalytic cracking process was developed by Eugene Houdry in the 1920s. This process was an outgrowth of his experiments on catalysts for removing sulfur from oil vapors.[1] These catalysts became deactivated due to the buildup of a carbonaceous deposit from the oils. Houdry discovered that this deposit could be burned off with air, and the catalyst activity restored. This discovery made a commercially viable process possible.

The Vacuum Oil Company (which later merged with Standard Oil of New York to form what was to become Mobil) supported Houdry in the development of the process. They were later joined by Sun Oil.[1,2]

After considerable development work, the first commercial Houdry unit came on stream in 1936. The yields from the process were vastly superior to those from competitive thermal cracking processes,[1] and catalytic cracking was quickly accepted. By 1943, 24 of these units were in operation or under construction. The combined capacity of these units was 330,000 barrels per day (BPD).[2]

The Houdry process used a cyclic, fixed bed configuration. Each unit consisted of several separate reactors. In operation, some reactors were in use while others were undergoing catalyst regeneration or gas purge. Each reactor was equipped with a molten salt heat removal system to remove the heat evolved during the regeneration step.[2] Heat removed during regeneration was transferred to the reaction step. The original catalyst used in the Houdry units were acid-treated bentonite clays.[3]

While the fixed bed catalytic cracking units were far superior to thermal processes, they were still limited by the fixed bed approach and the use of large catalyst particles.[1] This eventually lead to the development of two continuous processes, *moving bed catalytic cracking* (MBCC) and *fluid catalytic cracking* (FCC).

The moving bed process was developed by Houdry and the Socony-Vacuum Oil Company. This process addressed the limitations inherent in the fixed bed process by providing for continuous movement of catalyst from the reaction to regeneration[1]. Catalyst pellets were introduced at the top of the reactor vessel and flowed downward through the reaction zone. Vaporized feed was also introduced at the top of the reactor. Catalyst from the reactor section flowed to the regenerator (kiln) where they were contacted with air and the coke deposits burned off.

Early units used bucket elevators to move the catalyst to the top of the vessels. A later development introduced air lift.

The first of these units was a 500 BPD test unit in the Socony-Vacuum Paulsboro refinery. This unit was commissioned in 1941. A larger unit processing 10,000 barrels per stream day (BPSD) was commissioned in 1943 in Magnolia Oil's Beaumont, Texas refinery.[1]

Fluid catalytic cracking was developed as an outgrowth of work by Standard Oil of New Jersey (Exxon). In 1938, Catalytic Research Associates (CRA) was formed to develop catalytic cracking technology. The original members of CRA were: Standard of New Jersey (Exxon), Standard of Indiana (Amoco), M.W. Kellogg and I.G. Farben. The Texas Company (Texaco), Anglo Iranian Oil Company (BP), Royal Dutch Shell and Universal Oil Products (UOP) joined the group in 1940. I.G. Farben was dropped from the group at this time.[1]

Early efforts involved both pelleted and powdered catalyst. By mid-1940, however, the pelleted catalyst approach was abandoned. Initial work with powdered catalyst used long folded reactor lines. Screw conveyors were used to move the catalysts from areas of low pressure to areas of higher pressure.[1]

In 1940, researchers discovered that the catalyst could be made to flow down a vertical standpipe against a pressure gradient.[1] This eliminated the need for screw conveyors and greatly simplified the process.

The first FCC was commissioned in 1942 in Esso's Baton Rouge refinery. This unit used upflow reactors. Both the catalyst and air flowed upward through the reactor vessel and exited through the vessel overhead lines. External cyclones were used to collect the catalyst.

In this first unit, the feed was vaporized before being fed to the reactors. Heat was removed from the unit by catalyst coolers.[4] Later units charged liquid feed which was vaporized by the hot catalyst from the regenerator. This reduced or eliminated the need for external heat removal.

Even as the first unit was under construction, the next generation of FCC was in development. In this unit, the upflow reactors were replaced by downflow systems consisting of a dense fluid bed topped by a dilute phase. Internal cyclones were used to collect entrained catalyst and return it to the bed. The first of these Model II units came on stream in 1943.[1]

Developments and commercialization of both fluid catalytic cracking and moving bed cracking continued in parallel for some time. Eventually, however, the FCC process proved to be more flexible and came to dominate the field.

Catalyst developments have always played a major role in the development of catalytic cracking technology. The first catalysts were the acidized clays originally developed by Houdry. The first major advance was the development of synthetically prepared silica alumina catalyst in 1942.[1]

Early FCC units used both natural clay and synthetic catalysts. Both of these were prepared as powders by grinding[3]. Many of the particles in these ground catalysts were not spherical, but instead contained sharp corners and other irregular surface features. In operation, these irregularities would quickly break off leading to high attrition losses.

In 1948, the first spray-dried catalysts were introduced[1]. These catalysts were produced as microspherical particles and were produced with similar particle size distributions as the ground catalysts.[3] The spherical particles, however, showed both improved fluidization properties as well as a significant reduction in attrition related losses.

The early synthetic catalyst, contained approximately 13% alumina[3]. Early attempts to increase the activity by increasing the catalyst alumina content were not particularly successful and led to unacceptable yields of gas and coke. By 1954, however, these problems were solved and the first high alumina catalysts became available in 1955.[1,3] These catalysts contained 25–30% alumina.

Silica alumina catalysts dominated catalytic cracking. There was, however, some early work on silica magnesia blends.[3] These catalysts offered increased selectivity to middle distillate range materials. The interest in silica magnesia catalysts eventually faded, however, due to the drive for increased gasoline production.

The high alumina catalysts used in the 1950s and 60s were a significant improvement over previous natural and synthetic catalysts. By today's standards, however, they were low in activity and susceptible to rapid deactivation by coke deposition. Long feed/catalyst contact times were needed to reach the desired levels of conversion and units intended to operate with these catalysts were designed as bed crackers.

In these FCCs, the feed contacted the catalyst in a fluid bed. The catalyst in the bed was a mixture of fresh catalyst from the regenerator, partially spent catalyst that had been in the bed for a short time, and fully spent catalyst that had been in the bed for much longer.

This made it impossible to completely utilize the full activity and selectivity of the freshly regenerated catalyst. In addition, some of the regenerated catalyst moved quickly through the bed and back to the regenerator with little participation in the cracking reactions.

A watershed mark in fluid catalytic cracking occurred with the development of zeolitic catalysts by Plank and Rosinski of Mobil.

These first zeolitic catalysts showed significantly higher activities than the current high alumina catalysts. They also showed improved yield patterns and decreased deactivation by coke deposition over time.

The first commercial zeolite catalyst became available in 1964.[1] Using these catalysts it was possible to operate with increased conversion at shorter contact times and with lower catalyst to oil ratios.

Most bed cracking units utilized a catalyst transfer line or riser to move catalyst into the reactor bed. Feed and catalyst were mixed at the base of this line and the vaporized oil carried the catalyst up into the bed.

It was long recognized that some cracking occurred in this transfer line and that the yield selectivity for this cracking was superior to that in the reactor bed.[2] This is due to the fact that the feed oils see fresher, more active catalyst in the riser than in the back mixed bed.

The increased activity and selectivity of zeolitic catalyst made all riser cracking possible. The first all riser cracker was constructed by Shell in their Anacortes refinery. The improved yield structure possible from all riser cracking lead to its adoption as the standard for new unit design. Most bed crackers were converted to riser crackers either through internal modifications or by allowing the bed level to fall below the level of the riser exit.

In the early 1960s, M.W. Kellogg and Phillips Petroleum began the development of what was to eventually become the next frontier in catalytic cracking—*residual oil cracking*. The first purpose built resid cracker, or heavy oil cracker (HOC), was built in Phillips' Borger, Texas refinery.

The original concept was to operate the HOC as a feed preparation unit for the conventional gas oil cracker already in the refinery. The HOC would feed atmospheric bottoms and operate at low conversion. Heavy gas oils produced in the HOC would then be fed to the existing FCC where they would be converted further.

Kellogg and Phillips anticipated that the high carbon residue present in residual feeds would result in high coke yields and that some form of regenerator heat removal would be necessary. To meet this need the HOC was equipped with regenerator bed coils. These coils were submerged in the regenerator bed. During operation, boiler feed water was circulated through the coils. Steam generation in the coils absorbed heat from the regenerator bed.

On commissioning, the HOC produced an unexpected result. Conversion of the residual oil feed was in many ways comparable to the conversion of gas oil feeds in conventional FCCs. The yield pattern was, however, different. The high carbon content of the residual oil feed resulted in an elevated coke yield compared to gas oil units at similar conversions. In addition, the metals content of the resid resulted in high metals levels on the equilibrium catalyst inventory and this resulted in increased yields of light gases, especially hydrogen.

This first HOC was a great success and is still in operation today. Despite this, however, residual oil cracking did not gain significant interest until the mid-1970s. At that time, increases in the price of crude oil, decreasing demand for heavy fuel oil and a decrease in the availability of light crudes increased pressure on refineries to increase the yield of transportation fuels from each barrel of crude oil. This in turn led to renewed interest in cracking the "bottom" of the crude barrel.

In response to these pressures, two new residual oil cracking technologies—the UOP residual catalytic cracking (RCC) technology and the Total residual fluid catalytic cracking technology—were developed to compete with heavy oil cracking. Both of these technologies used two-stage regeneration to cope with the problem of excess coke production. Both were commercialized in the 1970s.

Today, FCC technology is available from a variety of licensors. The major licensors are:

Exxon
M.W. Kellogg
Stone & Webster/IFP (Total Technology)
ABB Lummus Global (Texaco Technology)
UOP

It is interesting to note that four of these five major licensors were members of CRA or rely on technology developed by an original CRA member. In addition, the other CRA members (Shell, Amoco, BP) have continued to develop FCC technology for their own use and/or for limited licensing.

Thus, the Stone & Webster/IFP joint licensing effort of the Total RFCC technology is the only significant new entry into the field of FCC development.

This fact means that most FCC technologies are the result of more than 50 years of continuing development involving both improved understanding of the chemical and physical processes involved as well as the equipment needed to control and direct the process to the desired goals. This has produced a technology that is mature in many ways but that continues to evolve to meet the changing needs of petroleum refining.

Role in Petroleum Refining

Fluid Catalytic Cracking is the primary conversion unit in many petroleum refineries. Crude oil, as produced from the

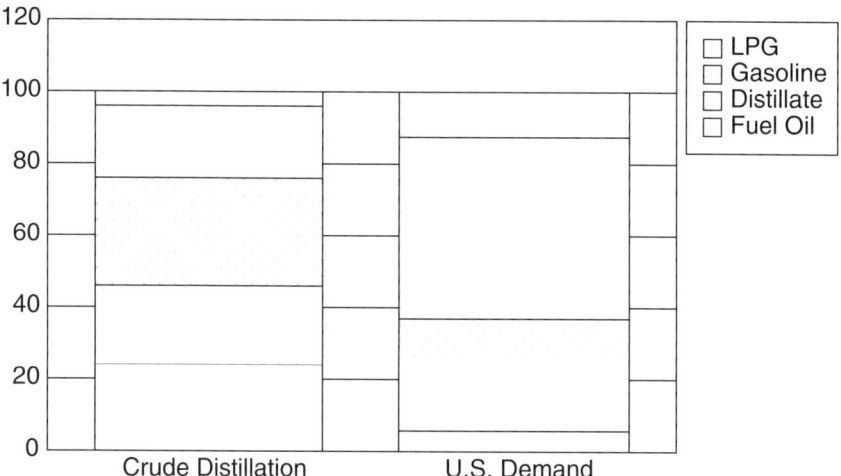

FIGURE 1-1. Products from crude.

ground, contains hydrocarbons ranging from light gases and LPG to resids boiling above 650°F (343°C). Products of various boiling ranges can be produced by distillation. Figure 1-1 shows the products available by distillation from a typical crude oil and a typical demand slate for petroleum products in the United States.[1] Compared to the products demand, crude oil is short of material boiling in the transportation fuel (gasoline and diesel) range and long on heavier material. Fluid Catalytic Cracking units convert a portion of this heavy material into lighter products, chiefly gasoline and middle distillates.

Figure 1-2 is a block flow diagram for a typical high conversion refinery. Crude oil is distilled in an atmospheric or crude distillation unit to produce LPG, naphthas, kerosene, and diesel oil. The residue from this distillation process—atmospheric residue or atmospheric tower bottoms (ATB)—is fed to the vacuum distillation unit where it is separated into vacuum gas oils and vacuum resid.

The heavy vacuum gas oil, which normally constitutes 25–35% of the total crude oil volume, is fed to the FCCU where it is converted to lighter products. In some refineries, the vacuum resid is fed to thermal conversion units such as a visbreaker or delayed coker. Where this is done, the heavy gas oils produced by these units may also be fed to the FCCU.

8 Chapter 1

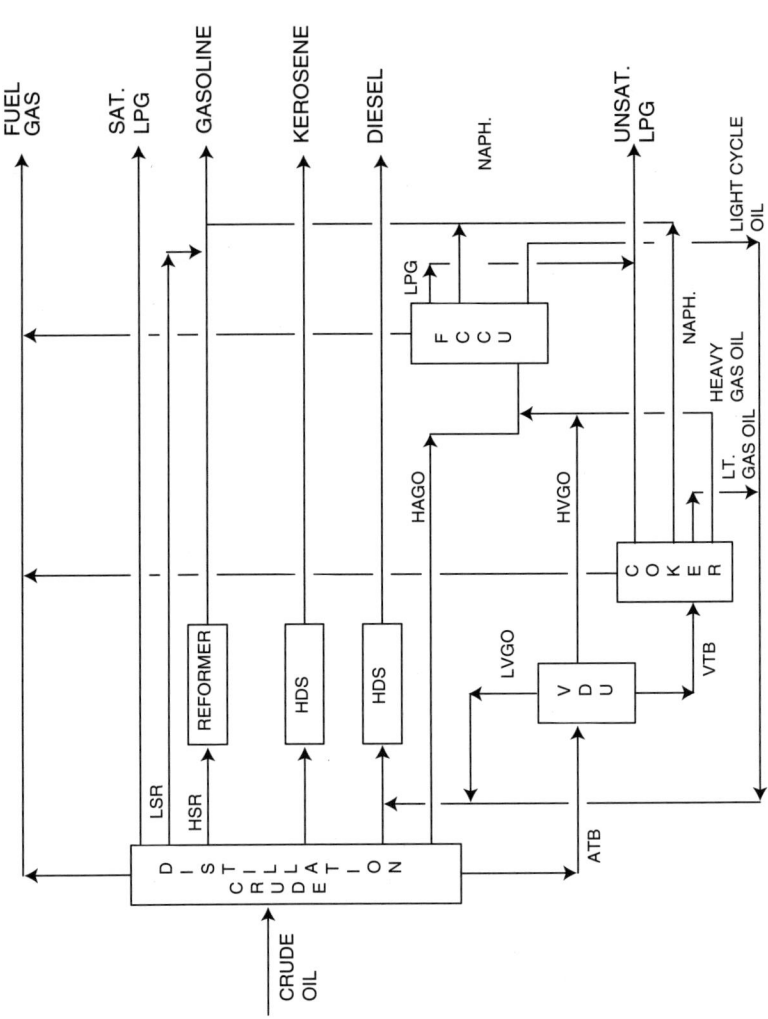

FIGURE 1-2. High conversion refinery.

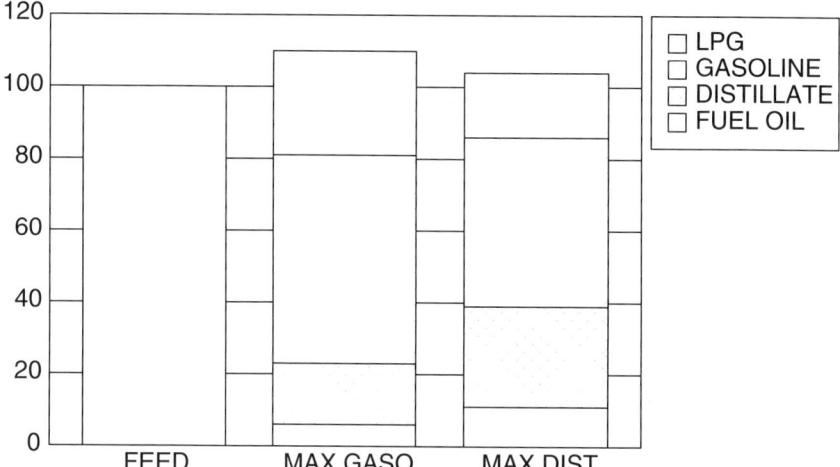

FIGURE 1-3. FCC feed and products.

The FCC can be operated to maximize the yield of gasoline or middle distillate (LCO). This flexibility allows the refiner to tune his product slate to changes in demand in his marketing area. Typical yield patterns from a gas oil cracker operating for maximum gasoline and maximum distillate are shown in Figure 1-3. The effect of this conversion on the product slate for a refinery are shown in Figure 1-4. As can be seen, these product slates are much closer to the product demand structures shown in Figure 1-1. By adjusting the operation of the FCC and varying the feed rate, the refiner can tune his individual product slate as necessary. Additional flexibility can be achieved by feeding atmospheric resid directly to the FCC. The result is an improved product slate and increased flexibility to adjust to changes in the market.

Process Description

A modern fluid catalytic cracking unit consists of three main sections: The reactor/regenerator, the main fractionator and the gas recovery section (also known as the gas concentration unit or the vapor recovery unit). In addition to these primary sections the feed is preheated in the feed preheat system, and the flue gas from the regenerator is processed in the flue gas system.

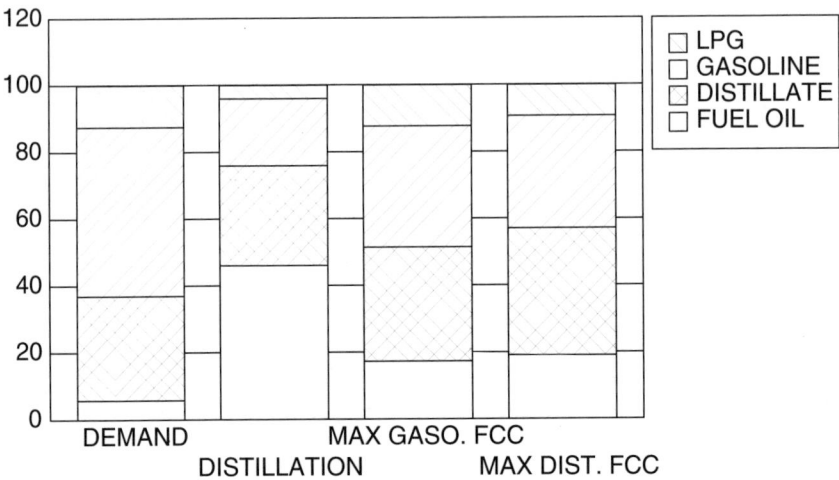

FIGURE 1-4. Effect of FCC on refinery yields (30% of crude fed to FCC).

Feed Preheat System

Oil feed to the reactor is first heated to the desired feed preheat temperature. This is usually done by heat exchange with intermediate heat removal pumparounds from the main fractionator. While feed preheat systems differ greatly from unit to unit, the feed will normally be heated by exchange with the light cycle oil, heavy cycle oil and/or bottoms pumparounds. Typically this will raise the feed temperature to 300–500°F. This is generally sufficient for most FCCUs. In some cases, however, a feed preheat furnace is included to further raise feed temperature prior to its injection into the riser.

Reactor/Regenerator System (Figure 1-5)

Following feed preheat, the hot feed is injected into the base of the riser where it contacts hot catalyst from the regenerator. Contact with the hot catalyst vaporizes the feed and the mixture of hot catalyst and oil vapors travels up the riser.

Process Overview 11

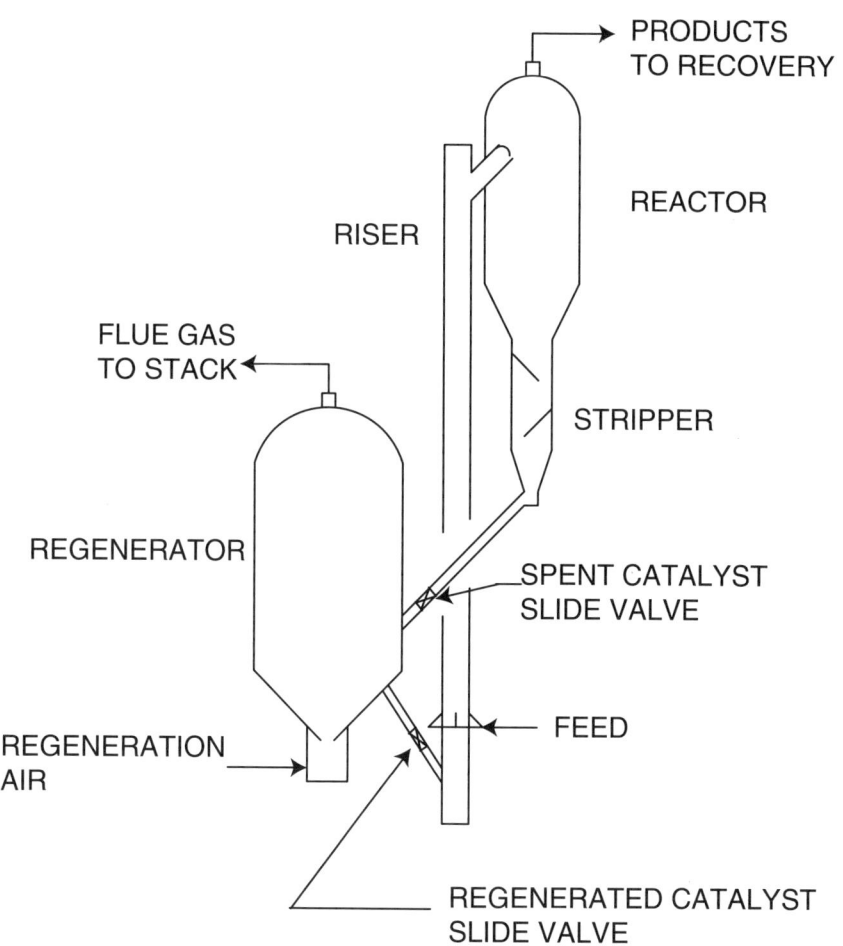

FIGURE 1-5. Reactor/regenerator.

Cracking reactions occur as the oil vapor and catalyst flow up the riser. Overall these reactions are endothermic and thus, the temperature in the riser decreases as the reaction progresses. Typically, the temperature will fall 40–70°F from the feed injection zone to the riser outlet. Residence times in the riser are typically 1–4 seconds, and the vast majority of the cracking reactions occur during this brief period.

At the end of the riser, the product vapors and the catalyst flow through a riser termination device (RTD) which separates the catalyst from the hydrocarbon vapors. A quick separation is essential since the "spent" catalyst leaving the riser actually retains considerable activity. If this catalyst were to remain in contact with the oil vapors, additional—and undesirable—reactions would occur.

Catalyst separated in the riser termination device is directed into the spent catalyst stripper. Hydrocarbon vapors from the RTD enter the reactor vessel.

In today's riser cracking units, the reactor vessel plays only a minor role in the actual cracking reactions. In fact, the reactions that do occur in this vessel are generally considered to be undesirable. The primary functions of the reactor vessel today are to provide some disengagement space between the riser termination device and the reactor cyclones and to contain both the RTD and the cyclones. Given this changed role, the term "reactor" is somewhat misleading. Some, in fact, prefer to use the term "disengager" when referring to this vessel.

The product vapors entering the reactor from the RTD are mixed with steam and hydrocarbon vapors leaving the spent catalyst stripper. This combined gas flow passes through the disengager and into the reactor cyclones. These may be single- or double-stage, but single-stage high efficiency systems are generally preferred for units with efficient riser termination devices.

The reactor cyclones remove any catalyst not separated by the RTD. This catalyst flows down the cyclone diplegs into the spent catalyst stripper. Vapors from the reactor cyclones flow through the reactor plenum and into the reactor overhead line.

Spent catalyst from the RTD and the reactor cyclones flows into the spent catalyst stripper. This catalyst still contains a considerable volume of product vapors. If not separated, these hydrocarbons will be carried into the regenerator where they will be burned. In the stripper, the catalyst is contacted with steam which displaces the hydrocarbon vapors. The bulk of the steam is injected at the bottom of the stripper and flows upward through the stripper while the spent catalyst flows downward. Most strippers are equipped with a series of baffles to improve the mixing between steam and catalyst.

Steam and stripped hydrocarbons flow out through the top of the stripper and mix with product vapors leaving the riser termination device. The stripped catalyst exits from the bottom of the stripper and enters the spent catalyst standpipe. Depending on the unit design, steam may be injected into the standpipe to improve the flow of catalyst.

At the bottom of the spent catalyst standpipe the catalyst flows through the spent catalyst control valve. This is normally a slide valve. This valve serves to control the flow of catalyst from the stripper and thus the stripper bed level. From the spent catalyst standpipe the catalyst flows into the regenerator.

In the regenerator the spent catalyst is contacted with air from the main air blower. The catalyst and air are well-mixed in a fluid bed or fast fluid bed and the carbon (coke) deposited on the catalyst during the cracking reaction is burned off. The heat produced by the combustion of the coke deposits raises the temperature of the catalyst by 300–400°F. Flue gases leaving the regenerator catalyst bed pass through the regenerator cyclones where entrained catalyst is removed and returned to the regenerator bed. Flue gases leaving the cyclones pass through the regenerator plenum and into the flue gas system.

Hot regenerated catalyst leaves the regenerator through the catalyst withdrawal hopper and flows into the regenerated catalyst standpipe. As with the spent catalyst standpipe, this standpipe may require aeration to assure smooth catalyst flow. Air is the preferred medium for aeration of regenerated catalyst standpipes.

Catalyst leaving the regenerated catalyst standpipe flows through the regenerated catalyst control valve and into the base of the riser where it contacts the fresh feed. The regenerated catalyst control valve controls the quantity of hot catalyst entering the riser and thus, the riser outlet temperature.

Flue Gas System

The regenerator flue gas passes through the flue gas slide valves, which control regenerator pressure, and into the flue gas line. In some cases, the flue gas flows directly to the stack where it is discharged into the atmosphere. In most units, however, the flue gas system contains additional equipment to recover energy (power recovery turbines, CO boilers and/or flue gas coolers) and to remove undesirable materials prior to discharge (third stage separators, electrostatic precipitators, flue gas scrubbers). These systems will be discussed in more detail in Chapter 5.

Main Fractionator (Figure 1-6)

Hydrocarbon products from the reactor cyclones flow through the reactor overhead line and into the bottom of the main fractionator. This tower is somewhat unusual since the tower feed is a superheated vapor which must be cooled before any liquid products can be condensed.

In the bottom of the main fractionator (the desuperheating zone), the hot vapors from the reactor are contacted by cooled circulating tower bottoms liquid; the vapors are cooled and the bottoms product is condensed. This part of the tower is normally equipped with either shed deck or disc and donut style baffles, but structured grids have also been used successfully.

Circulating liquid plus the net bottoms product are withdrawn from the bottom of the tower. This stream is cooled, usually by exchange with fresh feed and by steam production in the bottoms steam generator(s). The net product is separated from the circulating liquid stream, cooled further and sent to storage or other use. The circulating liquid stream is returned to the tower at the top of the desuperheating section.

Vapors from the desuperheating zone pass up through the tower where they are cooled further by circulating heavy cycle oil. This pumparound stream is normally cooled by heat exchange with fresh feed and reboilers in the gas concentration unit. The HCO pumparound may also be cooled by heat exchange with cold boiler feed water or by air or water trim coolers.

Following the HCO pumparound section, light cycle oil (LCO) is withdrawn from the tower and sent to the LCO stripper. In the LCO stripper, the product liquid is stripped by direct steam injection to control the quantity of low boiling hydrocarbons and thus, the product flash point. LCO from the bottom of the stripping tower is cooled and sent to storage.

Heat removal from the midsections of the main fractionator is by an LCO and/or heavy naphtha pumparound. The heat removed is used to preheat fresh feed, reboil towers in the gas concentration unit or to preheat boiler feed water. In some main fractionators, there is a heavy catalytic naphtha (HCN) side draw above the light cycle oil. In these towers, the HCN is usually stripped in a steam stripper for flash point control.

The overhead products from the main fractionator consist of cat naphtha (gasoline), C_3 and C_4 LPG and dry gas (C_2 and lighter

Process Overview **15**

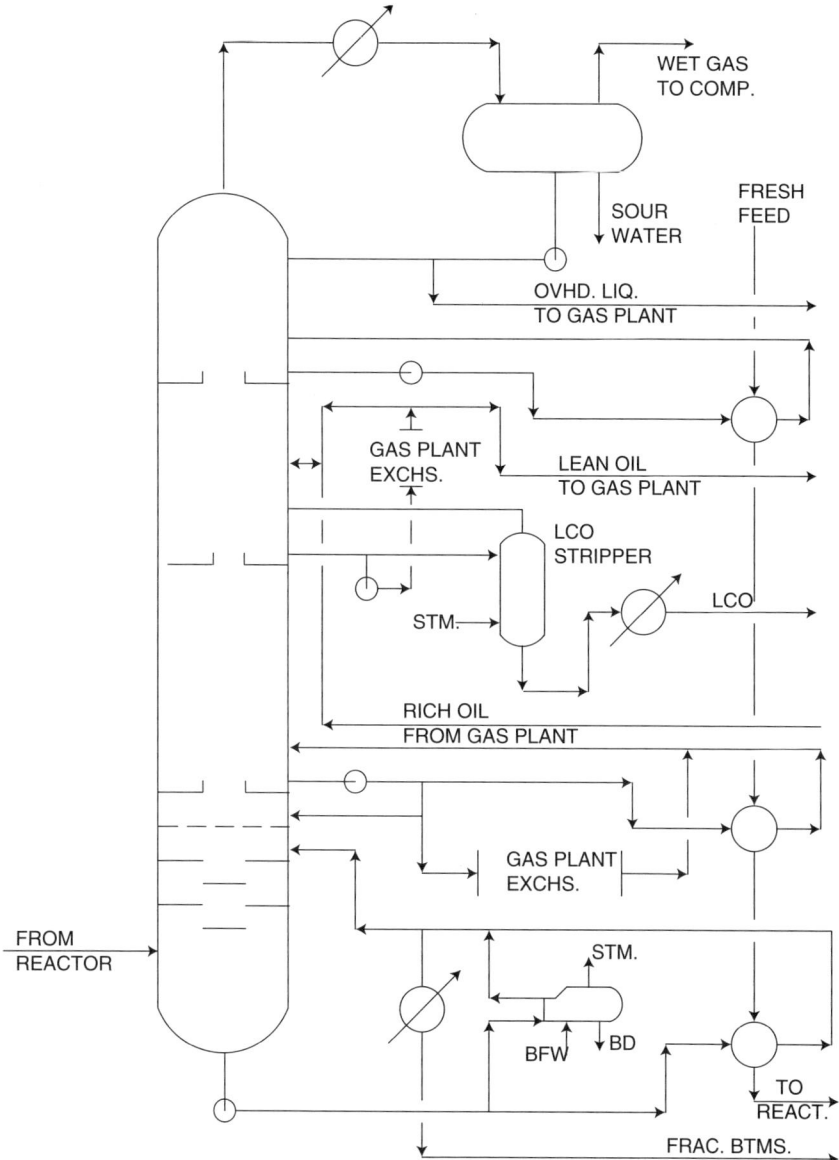

FIGURE 1-6. Main fractionator.

materials plus inerts carried into the reactor by the catalyst). This overhead gas is condensed in a partial condenser, and the gases and liquids are separated in the overhead separator drum. A portion of the condensed liquid is returned to the main fractionator as reflux, and the net overhead liquid is pumped to the gas concentration unit. Gases from the overhead drum flow to the wet gas compressor located in the gas concentration unit.

Gas Concentration Unit (Figure 1-7)

The gas concentration unit (GCU) provides recovery of the C_3 and C_4 LPG produced in the FCC and separation of the lighter liquid products. Gas from the main fractionator overhead drum flows to the wet gas compressor. This is usually a two-stage machine. The first stage discharge is cooled and partially condensed in the interstage cooler and the liquid and gas streams are separated in the interstage separator drum. The liquid is pumped to the high pressure condensers and the gas flows to the second stage of the gas compressor.

The gas discharge from the second stage of the gas compressor is combined with the primary absorber bottoms liquid, the stripper overhead vapors and the liquid from the compressor interstage drum. This combined stream flows through the high pressure condensers and into the high pressure separator. Gas from the high pressure separator flows to the primary absorber column.

Overhead liquid from the main fractionator is pumped to the primary absorber to serve as lean oil. In some cases, debutanizer bottoms liquid may also be pumped to the primary absorber to increase the lean oil rate and thus, the propylene reovery.

C_3 & C_4 LPG recovered in the primary absorber flow with the absorber bottoms to the high pressure condenser. Primary absorber overhead gas flows to the secondary or sponge absorber.

The sponge absorber is intended to capture gasoline range material (mostly C_5s) lost to the gas in the primary absorber. Lean oil is either unstripped light cycle oil or unstripped heavy naphtha from the main fractionator. The lean sponge oil is cooled, first by exchange with the sponge absorber bottoms and then by air or water coolers. Rich sponge oil from the absorber is returned to the main fractionator where the absorbed gases are vaporized and returned to the GCU by way of the gas compressor. The absorber overhead flows to treating and from there to the refinery fuel gas system.

Process Overview 17

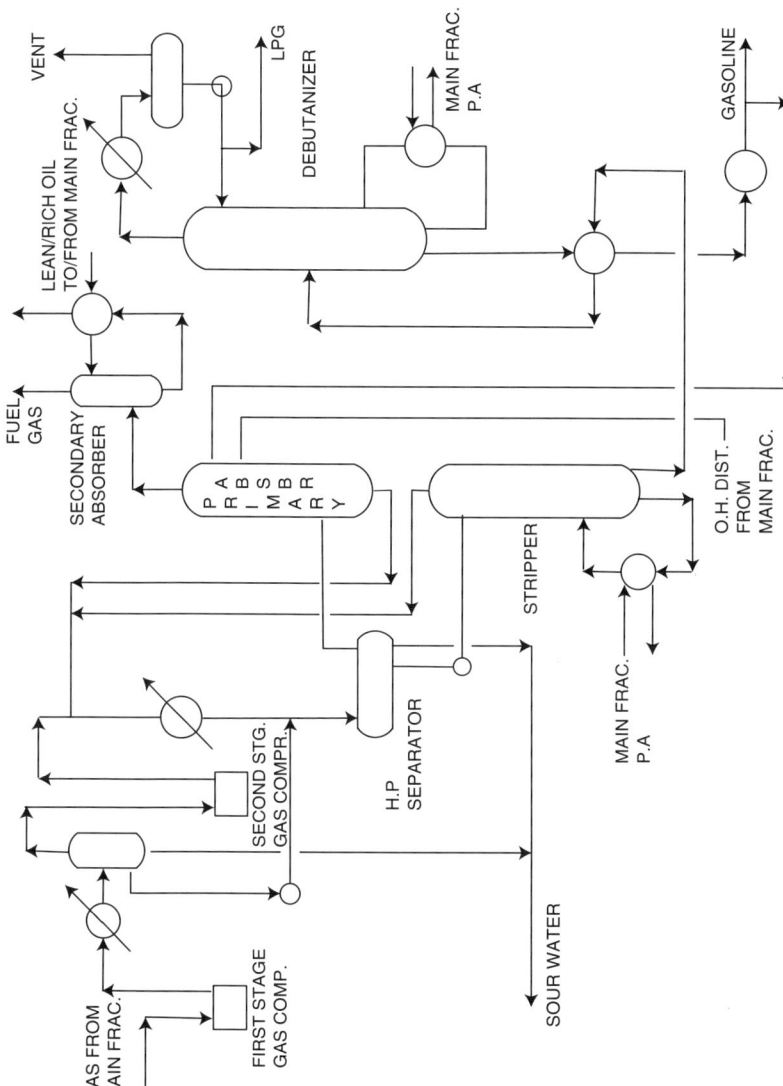

FIGURE 1-7. Gas concentration unit.

Liquid from the high pressure separator is pumped to the top of the stripper (de-ethanizer) tower. This tower removes ethane and lighter materials from the recovered liquid and is generally reboiled using heat from the main fractionator LCO or HCN pumparound. Gas leaving the top of the stripper flows to the high pressure condenser. The bottoms liquid, which contains the total liquid recovered in the GCU, is preheated by exchange with the debutanizer bottoms liquid and fed to the debutanizer column.

The debutanizer separates the recovered liquids into LPG and naphtha. This is a conventional distillation column. Heat for the reboiler is supplied by the HCO pumparound from the main column. The overhead vapors are totally condensed and supercooled in the overhead condenser. A part of the condensed liquid is returned to the tower as reflux and the remainder is yielded as LPG product.

Bottoms from the debutanizer are cooled by exchange with the tower feed and then by air or water coolers. A portion of the cooled bottoms stream may be returned to the primary absorber as *recycle lean oil* and the remainder is yielded as the net gasoline product from the GCU.

In some cases, the LPG from the overhead of the debutanizer flows to a depropanizer where it is separated into separate C_3 and C_4 products. Similarly, the gasoline product can also be split into light and heavy fractions.

LPG produced in the FCCU contains both hydrogen sulfide and mercaptan compounds. These must be removed. Typically, the LPG is treated in an amine absorption tower to remove the H_2S, and in a mercaptan extraction unit to remove the mercaptan sulfur.

The naphtha produced also contains mercaptans. These corrosive sulfur compounds are normally converted to disulfides in a sweeting unit.

Commercial FCC Designs

Since the first commercial FCC came on stream in 1942, there have been numerous designs used for the reactor/regenerator system. While all of these designs contain the same basic components (reactor, regenerator, catalyst transfer lines) individual configurations differ considerably. Each of these designs involve their own strengths and weaknesses which affect operation and reliability.

The following is a summary of the salient features of the common FCC designs. This section presents these designs in their original form. Many of the older designs have, however, been revamped to incorporate modern features.

Exxon (Esso) Designs

Model I. The first commercial FCC was an Esso Model I (Figure 1-8) upflow unit; few of this design were built and none remain in operation today. This design is included here because of its historical significance, and because some of its features can be found in more recent designs.

Feed to these units was first heated and vaporized in a feed preheat furnace. The vaporized feed was then mixed with hot catalyst from the regenerator and the mixture fed into the upflow reactor. The velocities in the reactor where high enough to form an upflowing or fast fluid bed. Catalyst and products left the top of the reactor vessel and flowed to external cyclones where the spent

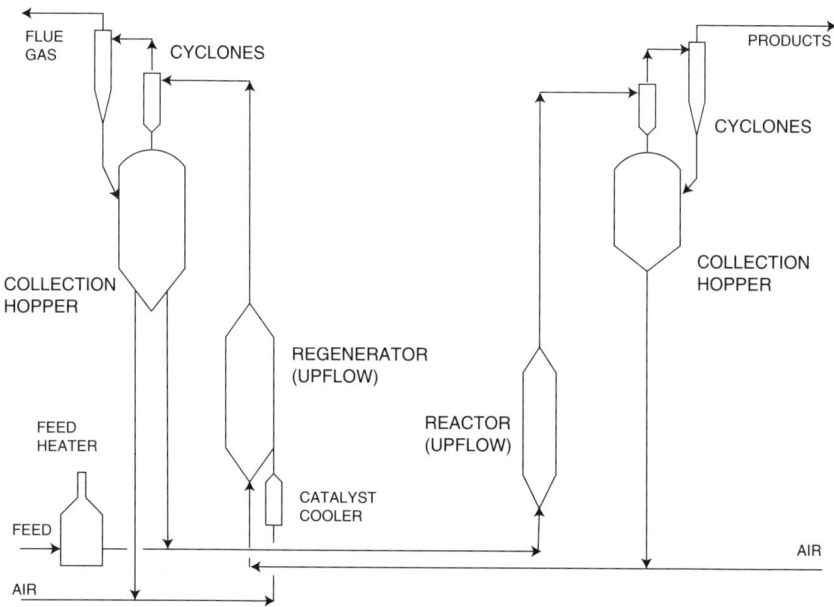

FIGURE 1-8. Esso Model I. (Adapted from: Reichie, A.D., *Oil and Gas Journal*, May 18, 1992.)

catalyst was separated from the products. The spent catalyst flowed into a holding vessel.

From this holding vessel the spent catalyst flowed through a slide valve and into the catalyst lift line. Regeneration air injected into the lift line transported the catalyst into the regenerator. As with the reactor, this vessel operated as a fast fluid bed. Regenerated catalyst and flue gas left the top of the vessel to the external regenerator cyclones. The flue gas from these cyclones passed through an electrostatic precipitator to recover addition catalyst and then to the stack. Catalyst from the precipitator and the regenerator cyclones flowed to the regenerated catalyst holding hopper. From this hopper, the regenerated catalyst flowed through a slide valve and into the fresh feed riser.

Catalyst was also withdrawn from the regenerated catalyst hopper and sent through catalyst coolers. Here the catalyst was cooled by generating steam. These coolers were necessary as the heat released in the regenerator was generally in excess of the heat required in the reactor.

Model II. The discovery that a fluid bed could be maintained at lower velocities lead to the development of the Esso Model II design (Figure 1-9). This was the first downflow design and it incorporates most of the features found in today's FCC units.

The Model II regenerator was elevated well above the reactor, and regenerated catalyst flowed downward though a very long standpipe. Lengths of 100–150 feet were not uncommon. At the base of this standpipe, the catalyst flowed through slide valves and into the short riser where it contacted the fresh feed. Unlike the Model I, feed to the Model II was in the liquid phase. This reduced the necessary feed preheat duty and also served to absorb some of the heat evolved during catalyst regeneration.

Model II units were designed as bed crackers and the short riser served primarily as a transfer line to move the catalyst and oil to the reactor vessel where it was distributed into the bed through a perforated grid. Spent catalyst from the reactor flowed through a short standpipe, slide valves and into the spent catalyst lift line. Regeneration air, introduced at the base of the lift line carried the spent catalyst up the lift line, through the regenerator grid and into the regenerator bed.

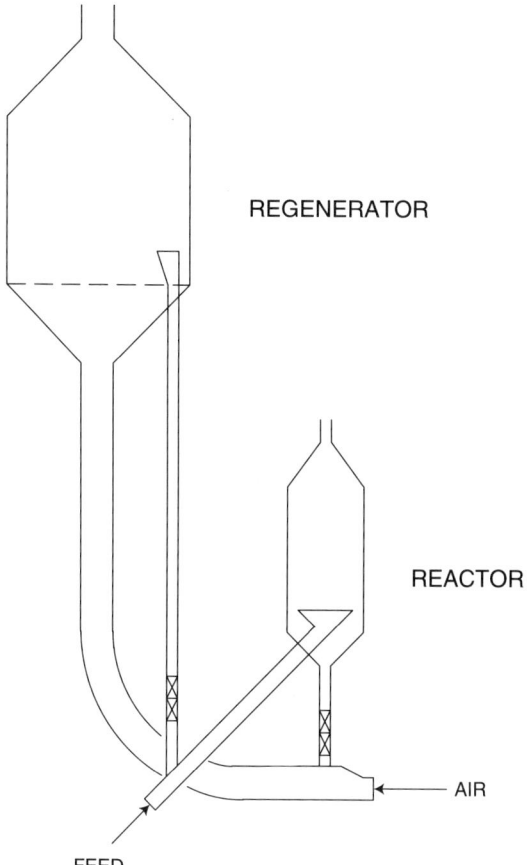

FIGURE 1-9. Esso Model II.

The long regenerated standpipe on the Model II unit resulted in very high pressure drops across the regenerated catalyst slide valves. This resulted in rapid erosion of these valves; it was standard practice to install two, or even three, valves in series on these standpipes. Modern slide valve designs have largely eliminated these problems. The long standpipes were often difficult to operate and subject to flow interruptions. Since both catalyst and air flowed through the grid, it was subject to erosion. In addition, the high velocities in the grid increased the attrition of the catalyst.

Model II units have proved to be a durable design. Several of these units—revamped to riser cracking and with improved catalyst and air distributors—are in operation today.

Model III. Experience with the Model II units showed that the regenerated catalyst standpipes could be shortened considerably. This resulted in the Model III unit (Figure 1-10). While this is technically an Esso design, much of the development work was done by M.W. Kellogg. Due to this fact, these units are often referred to as Kellogg side-by-side units.

The flow in these units is essentially the same as that in the Model II design. The reactor and regenerator are, however, at approximately the same elevation and the regenerated catalyst standpipes are shorter. This improved operation somewhat. As with the Model II, catalyst was distributed into the fluid beds through perforated grids which were subject to erosion. Many of these units are also still in service.

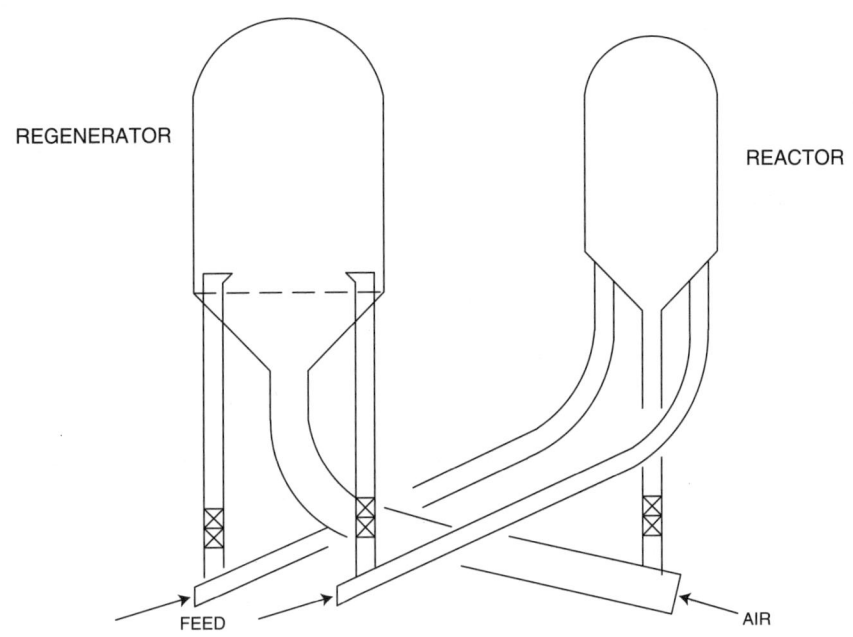

FIGURE 1-10. Esso Model III.

Model IV. The Esso Model IV reflected a departure from earlier designs. The slide valves used for catalyst control in the Model II and Model III units were subject to rapid erosion and were a major maintenance problem on these units. The Model IV concept attempted to eliminate this problem by eliminating the catalyst control valves.

Figure 1-11 is a schematic of a Model IV unit. The reactor and regenerator are at essentially the same elevation. Catalyst leaves the regenerator by overflowing into the overflow well. From the overflow well the catalyst flows down the regenerated catalyst standpipe, through the regenerated catalyst U-bend and into the riser. Feed is injected at the base of the riser, and the oil catalyst mix flows up the riser, through the reactor grid and into the reactor bed.

Spent catalyst from the reactor flows down through the spent catalyst stripper, into the spent catalyst U-bend and into the spent

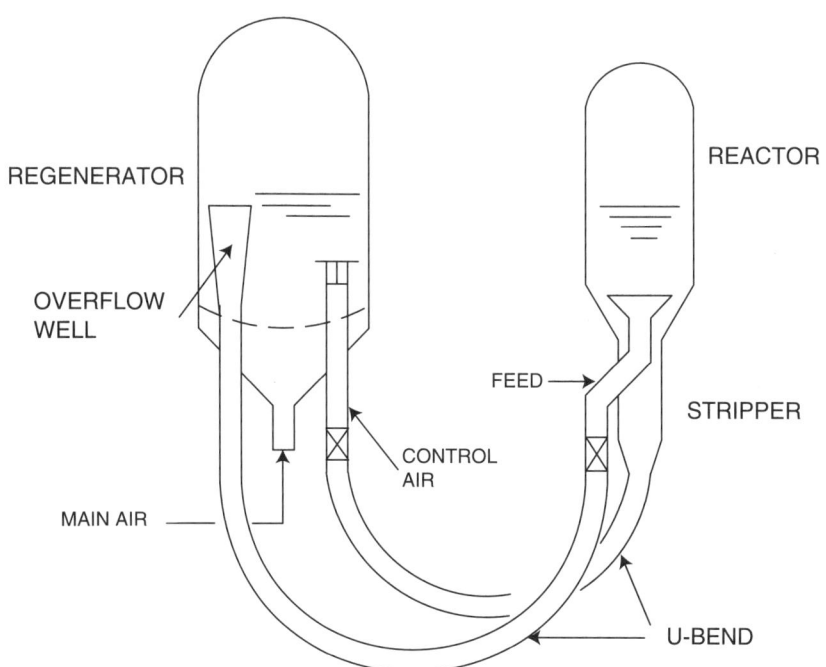

FIGURE 1-11. Esso Model IV.

catalyst riser where it is transported to the regenerator by the injection of "control" air. Additional combustion air is injected into the regenerator bed through the regenerator air grid. Catalyst circulation was controlled by changing the pressure differential between the reactor and the regenerator or by adjusting the control air rate. The slide valves shown in Figure 1-11 are used only as isolation valves during start up and shut down.

The Model IV configuration was fairly well-received and several of these units were constructed. The fact that these units did not require catalyst control valves eliminated a maintenance problem. In addition, the fact that spent catalyst did not flow through the regenerator grid reduced erosion and maintenance on this item as well.

In actual operation, the catalyst circulation is usually controlled by adjusting the pressure difference between the reactor and the regenerator. This changes the allocation of the catalyst inventory between the reactor and the overflow well. In addition to this pressure difference, changes in air rate, stripping steam rate and catalyst inventory all affect catalyst circulation and thus reactor temperature.

The catalyst U-bends can be a source of operational difficulties. At the bottom of the U-bend, the catalyst is essentially flowing horizontally. At this point, there is a tendency for the catalyst to defluidize at the bottom of the U-bend. To counter this tendency, U-bends are fitted with an extensive fluidization system. Proper operation and maintenance of this system is essential to maintaining smooth operation.

In most Model IV units, the reactor temperature is not controlled automatically. Thus, the units require more supervision than slide valve controlled units. Some Model IVs have, however, been modified so that the reactor temperature is used to control the reactor/regenerator pressure differential.

Exxon Flexicracker. The Exxon Flexicracker (Figure 1-12) design retains the overflow well used in the Model IV and thus, there is no regenerated catalyst slide valve. The difficult-to-operate U-bends are replaced, however, by a standpipe followed by upwardly sloped laterals. This combination is generally

Process Overview **25**

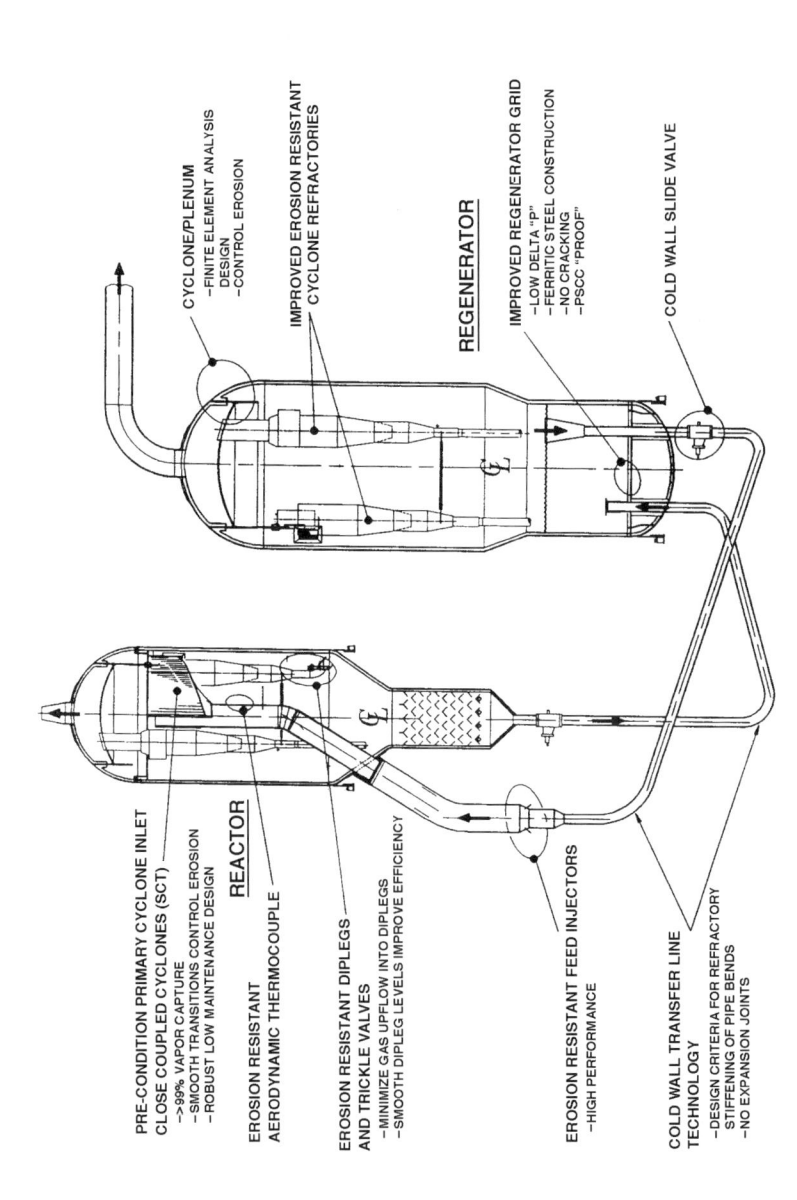

FIGURE 1-12. Exxon Flexicracker. (Shaw, D.F. et al., "FCC Reliability Mechanical Integrity," NPRA Annual Meeting Paper AM 96–24 (1996)).

referred to as a J-bend. Flow in these transfer lines is considerably easier to manage than in the earlier U-bends.

Flow from the stripper into the spent catalyst J-bend is controlled by a slide valve. Reactor temperature is controlled by adjusting the pressure differential between the reactor and the regenerator. Modern Flexicracker designs are equipped with short contact time RTDs and efficient feed nozzles. Combustion air is distributed in the regenerator by an improved plate grid which eliminates many of the problems associated with earlier grid designs.

M.W. Kellogg

Orthoflow A. The first stand-alone Kellogg design was the Orthoflow A (Figure 1-13). This was a stacked unit (a feature that Kellogg has retained to this day) with the reactor located above the regenerator vessel. Catalyst flow was regulated by plug valves instead of slide valves. This and the stacked configuration allowed Kellogg to design the unit with all vertical flow (hence the name orthoflow). Catalyst flowed from the regenerator directly into the base of the riser; there was no regenerated catalyst standpipe. Oil was injected into the riser through the hollow stem of the regenerated catalyst plug valve.

The spent catalyst stripper was internal to the reactor vessel. Both the riser and the spent catalyst standpipe were internal to the regenerator. The relatively short, vertical standpipe made these units easy to operate.

The common head between the reactor and the regenerator and the internal risers were all designed for the low operating temperatures of the time. As a result, these units frequently require materials upgrading to operate at today's conditions. The early plug valve designs were subject to the same erosion problems experienced by slide valves. As with slide valves, these problems have been eliminated through improved design.

Orthoflow B. Like the Esso Model IV, the Orthoflow B was a departure from conventional FCC design practices. As with all Orthoflow designs, these were stacked units. In the Orthoflow B, however, the reactor was on the bottom (Figure 1-14). This unusual arrangement was intended to reduce the discharge pressure of the main air blower, and thus reduce energy consumption.

Process Overview **27**

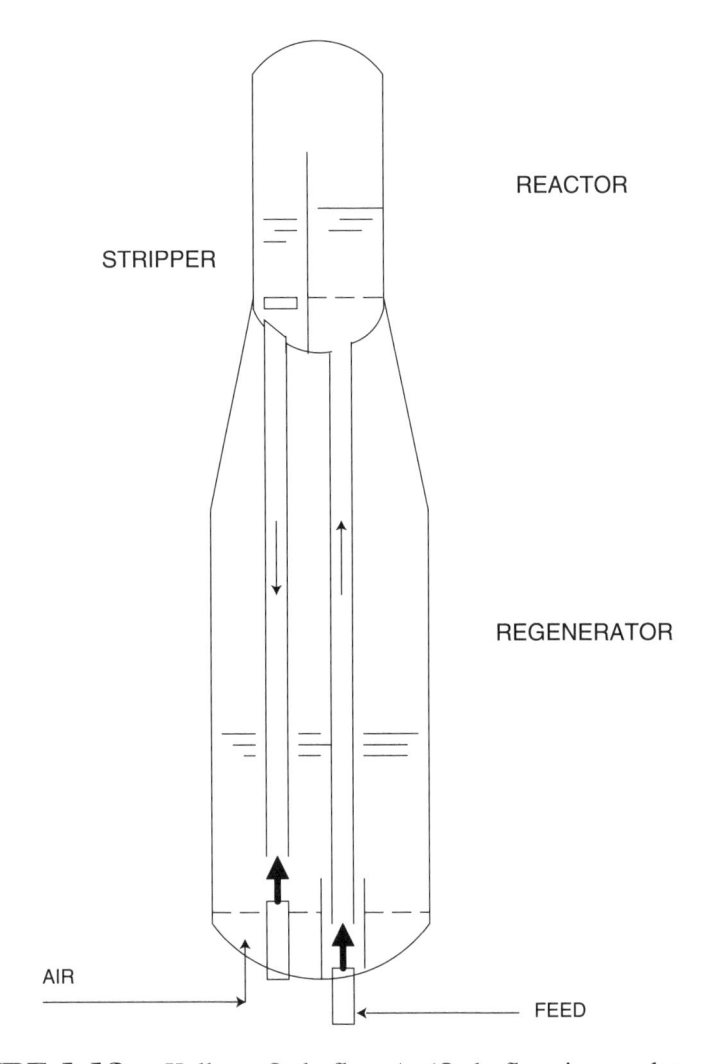

FIGURE 1-13. Kellogg Orthoflow A. (Orthoflow is a trademark of M. W. Kellogg.)

Regenerated catalyst flowed from the regenerator to the reactor through two vertical standpipes. Plug valves were used to control the catalyst flow and thus the reactor temperature. Catalyst flowed from the standpipes directly into the reactor bed; there was no riser. Feed was injected into the reactor through a number of pipe nozzles in the bottom head. Thus, the Orthoflow B was the

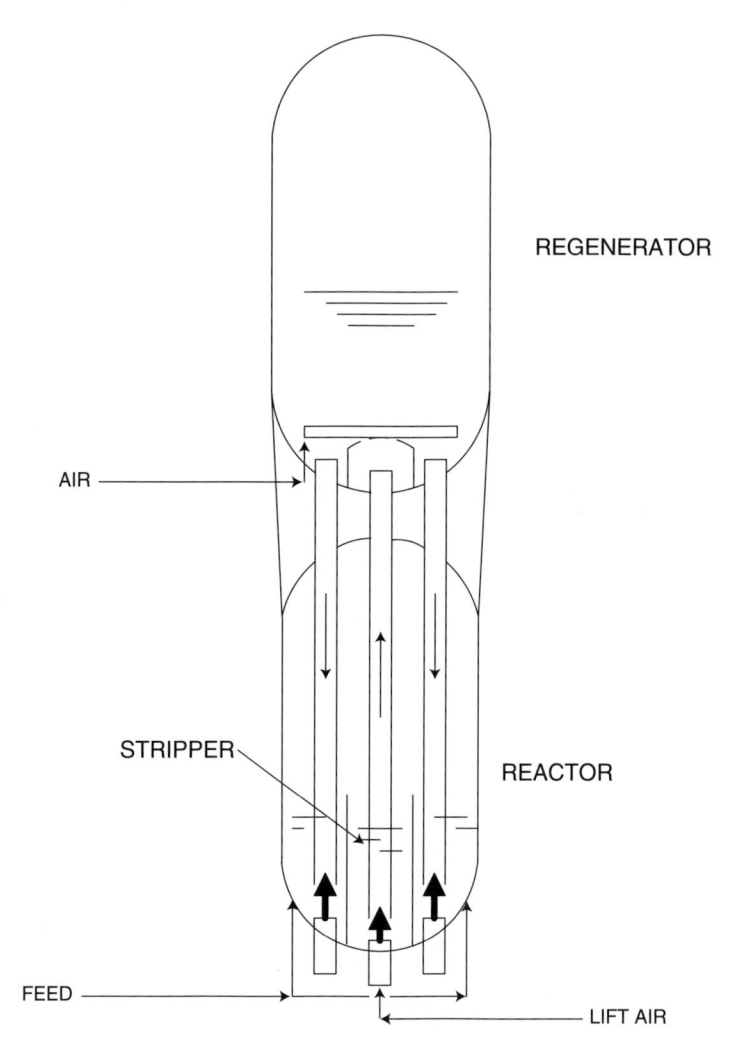

FIGURE 1-14. Kellogg Orthoflow B.

only true bed cracker ever built. Spent catalyst from the reactor was lifted to the regenerator using combustion air injected into the lift line through a hollow stem plug valve.

The Orthoflow B worked very well as a bed cracker. Since the feed was injected directly into the bed, it was never in contact with hot catalyst from the regenerator. This reduced thermal cracking reactions and gave improved yields.

Some Orthoflow Bs were constructed with a common head between the reactor and the regenerator. This can present mechanical problems at high regenerator temperatures.

Orthoflow C. The Orthoflow C marked a return to a more conventional configuration with the reactor on top and the regenerator on the bottom (Figure 1-15). Most Orthoflow C units had

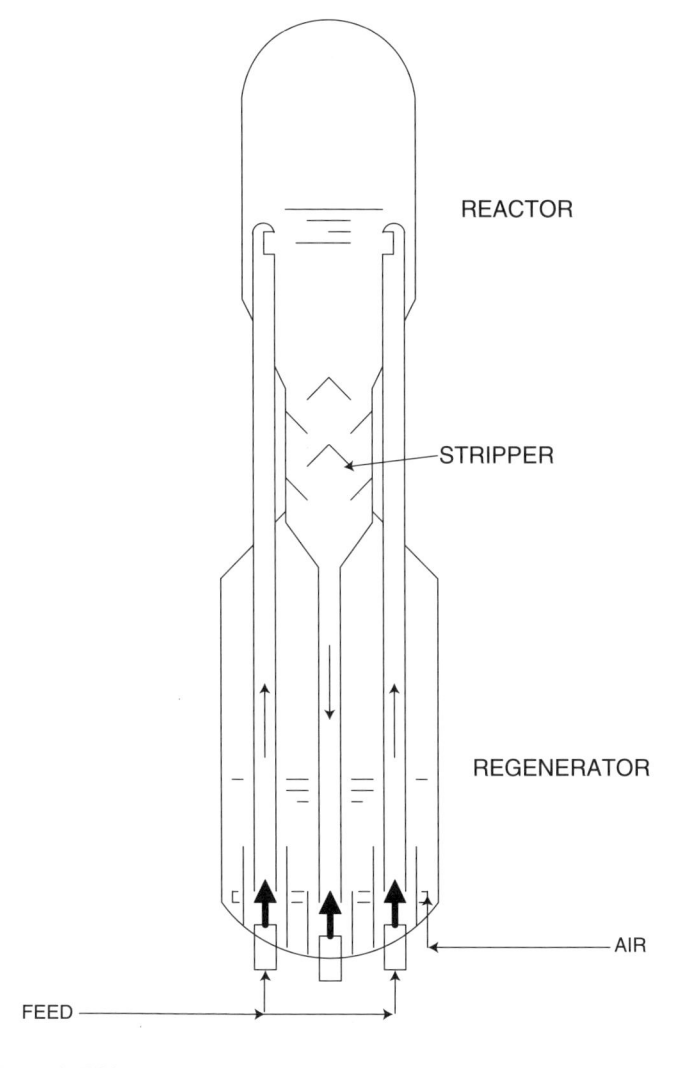

FIGURE 1-15. Kellogg Orthoflow C.

two risers, one riser for fresh feed and one riser for recycle. Both risers passed through the top head of the regenerator and entered the bottom of the reactor vessel. As with the Orthoflow A, catalyst flowed directly from the regenerator bed into the riser and there were no regenerated catalyst standpipes.

Early Orthoflow Cs were designed as bed crackers and the risers terminated low in the regenerator bed. Later versions were designed as riser crackers and the risers extended into the reactor and were equipped with simple inertial termination devices. The Orthoflow C configuration holds the distinction of being the first FCC design used in a purpose built resid cracker.

Orthoflow F and UltraOrthoflow Designs. The Orthoflow F and UltraOrthoflow units are essentially the same in overall concept. The UltraOrthoflow design was the result of a joint licensing agreement between M.W. Kellogg and Amoco (no longer in effect) that incorporated Amoco's UltraCat regeneration technology into the Orthoflow design.

The Orthoflow F retained the stacked design of the A and C models. The riser was, however, moved outside of the reactor and regenerator vessels. This eliminated the temperature limitations of the earlier designs and allowed the riser to be fabricated of carbon steel with internal insulation.

Early Orthoflow F designs used a simple inertial separator on the riser termination. Since the mid-1970s, however, the standard design has been to use riser cyclones (sometimes called rough cut cyclones) as the RTD. Since the late 1980s Kellogg has offered Mobil's close coupled cyclone technology as a part of the Orthoflow design. The Orthoflow F also incorporated an improved plug valve design that eliminated erosion and sticking problems of earlier units. Other Orthoflow F features included multiple feed injection nozzles and an external plenum for the regenerator cyclones.

Some early Orthoflow F designs included the first widespread commercial use of two-stage regeneration. In these units, the regenerator was divided by an internal baffle.

The Orthoflow design includes a sophisticated spent catalyst distributor to spread catalyst over the regenerator cross section. Air to the regenerator is distributed through pipe grid distributors. Figure 1-16 shows a modern Orthoflow design.

Process Overview **31**

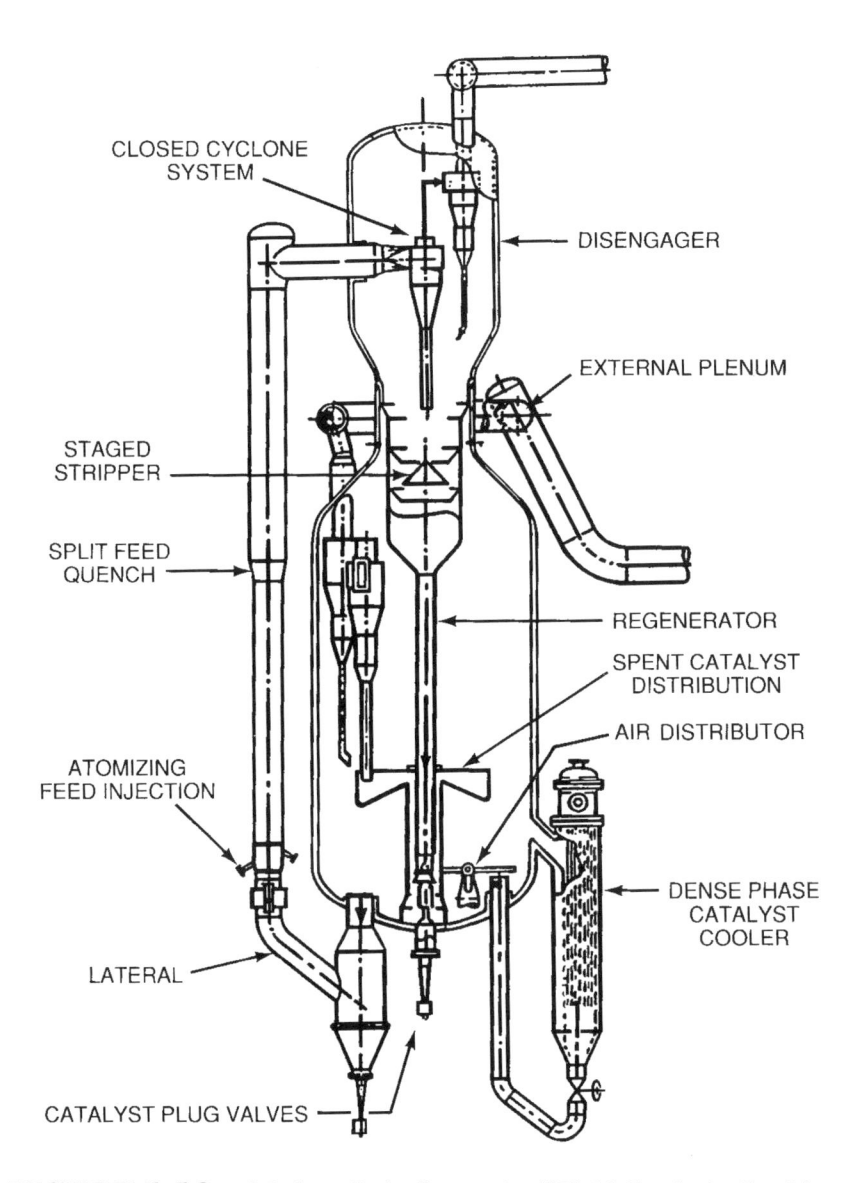

FIGURE 1-16. Modern Orthoflow unit. ("Fluid Catalytic Cracking: The Heart of a Refinery," M.W. Kellogg, 1993.)

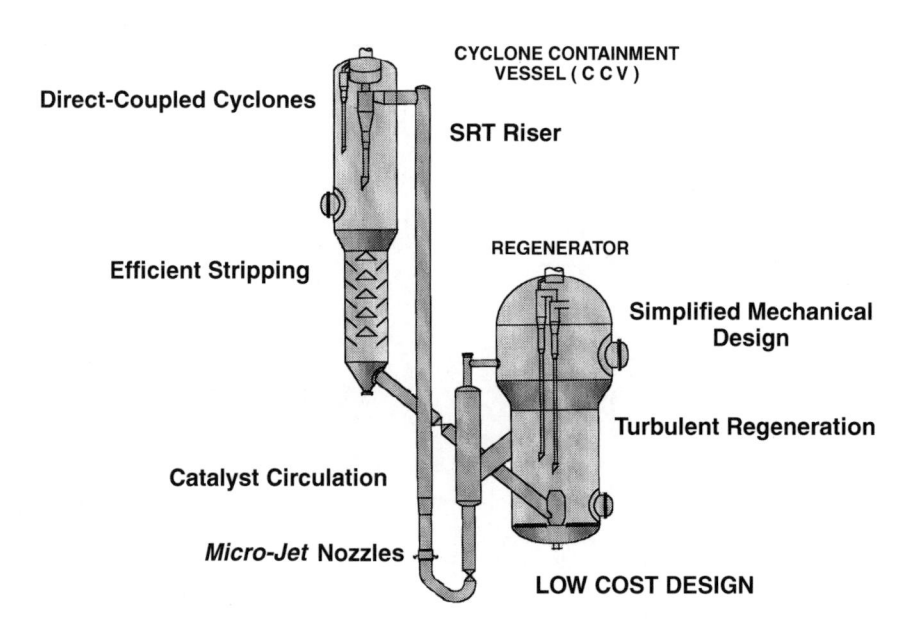

FIGURE 1-17. Lummus FCC. (Drawing courtesy of ABB Lummus Global.)

Lummus (Texaco)

Early Texaco designs used a curved sweeping riser that terminated in a "dragon head" inertial separator. These units were designed to operate with a catalyst bed in the reactor and were equipped with a slide valve to control the flow of catalyst from the reactor to the stripper.

Modern designs (Figure 1-17) use a vertical external riser, multiple feed injection nozzles and closed cyclone riser termination. The regenerator is unusual in that it makes use of a relatively high velocity turbulent bed followed by an expanded dilute phase section. This gives improved mixing of the air and catalyst. In 1996 Texaco sold their FCC technology to Lummus.

UOP

Stacked Unit. The UOP stacked unit (Figure 1-18) was a contemporary of the Esso Model III unit and the Orthoflow A. As with

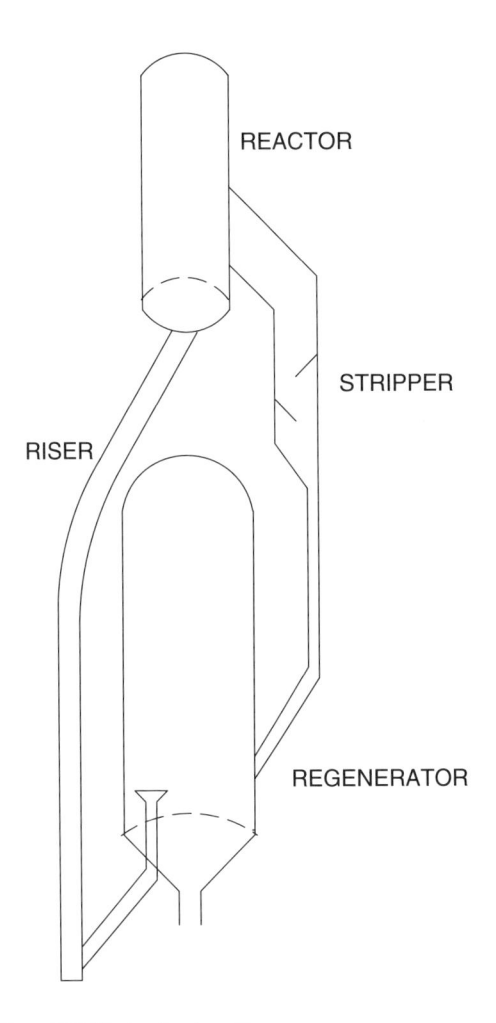

FIGURE 1-18. UOP stacked unit.

the Orthoflow A it was a stacked unit with the reactor on top of the regenerator. Unlike the Orthoflow A, however, the riser and spent catalyst standpipe were external. These units were designed as bed crackers and the riser discharged through a grid into the reactor bed. The spent catalyst stripper was located off the side of the reactor. The UOP stacked unit was easy to convert to all riser cracking and many of these units are still in operation today.

Semi-Stacked (Side-by-Side) Unit. The semi-stacked or side-by-side design (Figure 1-19) was used for larger units where mechanical considerations made the stacked design difficult. This design offered an all vertical riser and thus, it became the standard for UOP riser crackers. The semi-stacked design also has very short standpipes and this greatly simplified the design of the catalyst circulation system. Unlike units with long vertical standpipes, circulation in these units is not sensitive to changes in catalyst properties.

Due to the fact that the catalyst enters the side of the regenerator, these units generally show a temperature difference across

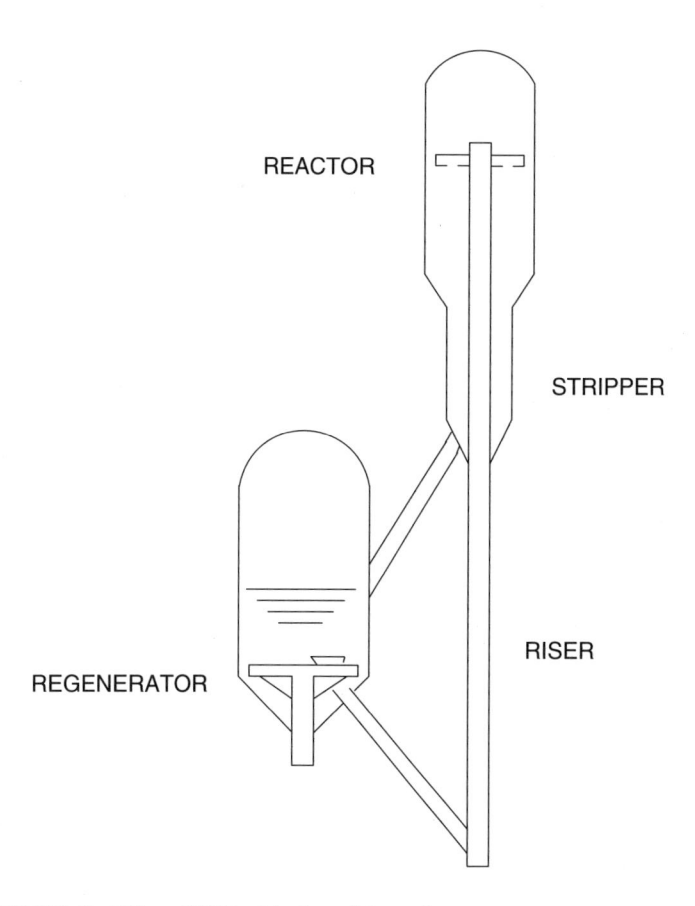

FIGURE 1-19. UOP side-by-side unit.

the regenerator cross section. In some cases, this can result in operating limitations.

High Efficiency Regenerator. The high efficiency regenerator design (Figure 1-20) uses a high velocity regenerator in place of the traditional fluid bed. Velocities in the combustor produce a fast fluid bed and the catalyst and air both travel upward into the disengaging vessel above the combustor. The high velocities in the combustor also result in very good mixing of the spent catalyst and air and thus, improve the burning kinetics in the

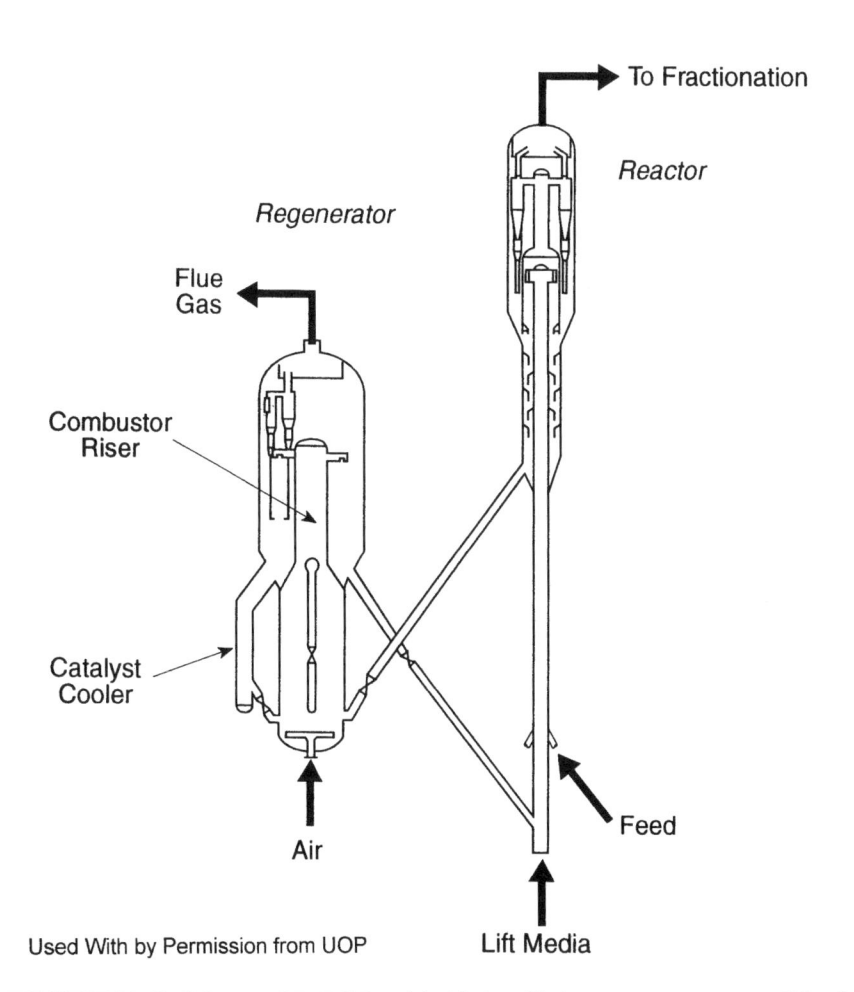

FIGURE 1-20. UOP FCC with high efficiency regenerator. (Used with permission from UOP.)

regenerator. This allows for a lower regenerator volume and thus a greatly reduced catalyst inventory.

This design has no significant weaknesses. In modern designs, the high efficiency regenerator is coupled with UOP's proprietary feed nozzle design and short contact time riser termination technology.

Stone & Webster/IFP. Stone & Webster and IFP jointly license FCC technology originally developed by Total Petroleum. This technology represents the only entry in the FCC field not associated with the original Catalytic Research Associates group.

The Total FCC design was originally developed for residue cracking. However, many of these units process conventional gas oil feeds.

The primary feature of this technology is a two stage regenerator (Figure 1-21). The first stage, located at the bottom of the stacked regenerator system operates in an oxygen deficient mode so that only a fraction of the carbon is burned to carbon dioxide. Normally 60–70% of the coke is burned in the first stage.

The partially regenerated catalyst from the first stage is lifted into the second stage using a portion of the second stage combustion air. The flow of catalyst between regenerator stages is controlled by a plug valve.

Catalyst regeneration is completed in the second stage which operates with sufficient excess oxygen to insure that the carbon on the catalyst is fully burned to CO_2. Regenerated catalyst from this stage exists from the side of the regenerator and flows into a fluidized catalyst hopper. From here it flows down the regenerated catalyst standpipe and into the base of the riser. The second stage regenerator uses external cyclones and this is probably the most obvious feature of these units.

Operation of the riser/reactor system is conventional. Feed is injected into the riser through multiple atomizing nozzles. This technology was the first to use high efficiency atomizing nozzles for feed injection. The riser is equipped with an efficient riser terminator.

Stone & Webster/IFP units have been well-received and there are many of these units in operation.

Early designs used an unbaffled stripper which did not perform as well as baffled strippers offered by other licensors. Recent designs have included a baffled stripper.

Process Overview **37**

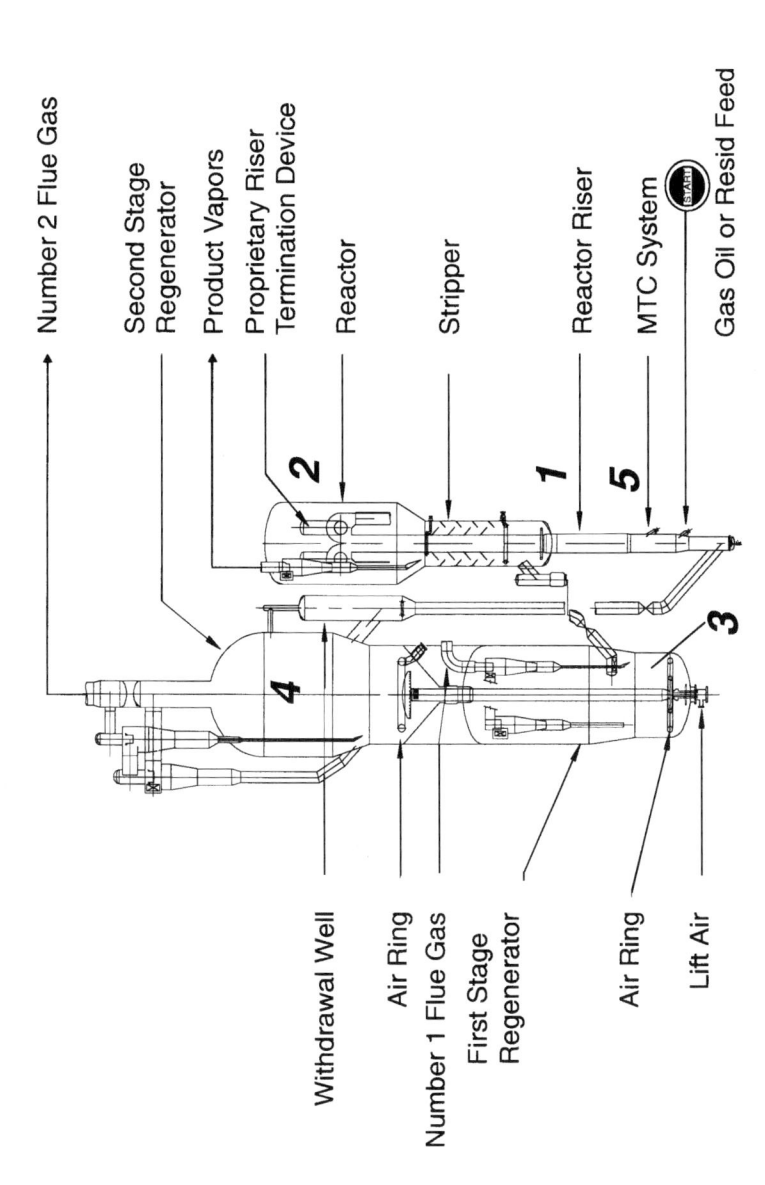

FIGURE 1-21. Stone & Webster/IFP unit with two stage regeneration. (Drawing courtesy of Stone & Webster Engineering Corporation, ©1994, Stone and Webster Technology Corp.)

Since the flue gas from the first stage regenerator contains considerable carbon monoxide, some form of CO combustion in the flue gas may be required. This adds to the overall cost of the unit. In addition, the relatively low overall level of heat released from the coke burned in the regenerator can limit the processing of light, easily cracked feeds.

Many of the features used in the Stone & Webster/IFP design are also available with a more conventional single stage regenerator (Figure 1-22). This configuration would be used for units designed to process light feeds or in revamps of existing FCCUs.

Shell

Shell was one of the original members of CRA. Over the years, they have continued to develop FCC technology for their own use. Some of these developments have been made available to others.

Shell was the first to use all riser cracking. Shell technology has traditionally stressed innovative cyclone and riser separator design as well as efficient stripping.

Residual Catalytic Cracking Units

The first purpose built residual catalytic cracking unit was a joint development between M.W. Kellogg and Phillips Petroleum. This unit was brought onstream in 1961. Essentially an Orthoflow C, this first heavy oil cracking unit, or HOC, used steam coils in the regenerator bed to remove the excess heat of regeneration produced by residual oil cracking. This allowed the unit to operate at high conversions.

Since the development of this innovative technology, Kellogg has continued to offer heavy oil cracking units based on the latest Orthoflow configurations. Modern designs incorporate external dense phase catalyst coolers to remove the excess heat of regeneration.

As discussed above, the Total FCC technology offered by Stone & Webster/IFP was developed for residual cracking. The original concept was to minimize the heat of regeneration by burning much of the coke under conditions that favored the production of carbon monoxide. In addition, the second stage regenerator was allowed to reach very high temperatures. This essen-

Process Overview **39**

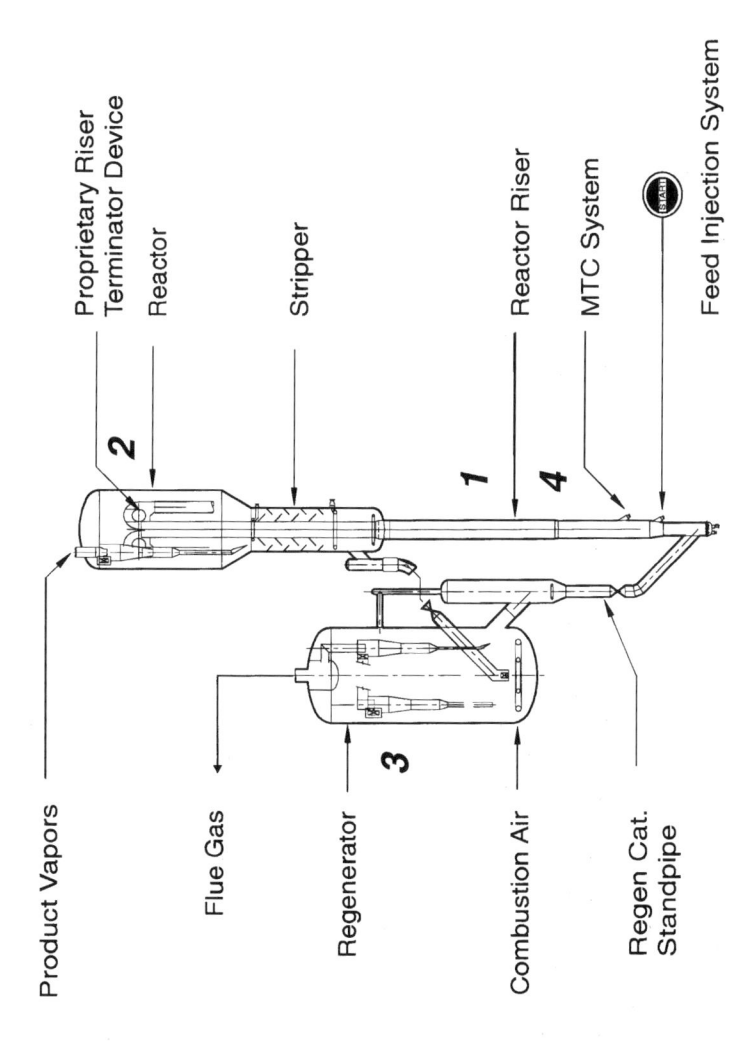

FIGURE 1-22. Stone & Webster/IFP unit with single stage regeneration. (Drawing courtesy of Stone & Webster Engineering Corporation, ©1994, Stone and Webster Technology Corp.)

tially forced the unit into low catalyst circulation rates and thus low conversions.

To maintain reasonable conversions, it was necessary to limit the second stage regenerator temperature to the conventional 1300–1400°F (704–760°C). This limited feed stocks to resids with low coke-making tendencies. To address this problem, Stone & Webster began offering catalyst cooler equipped units in 1990.

UOP working with Ashland Petroleum developed the RCC process in the 1970s. A typical RCC converter is shown in Figure 1-23. As with the Total technology, the RCC uses two stage regeneration to minimize the heat released by coke combustion. The RCC regenerators are stacked with the first stage on top of the second stage. The second stage regenerator does not have any cyclones, and flue gas from this stage passes into the bed of the first stage. Thus, there is only one flue gas system.

Normally 70 percent of the total coke produced is burned in the first stage. Catalyst coolers were a part of the first RCC design and continue to be an integral part of the technology.

New Developments

UOP has recently begun to market two new concepts in catalytic cracking. The first of these, millisecond catalytic cracking or MSCC was developed by BAR-CO Industries.

This process is designed to operate at very short contact times.

The advantages of short contact times are well-known. The product selectivity of freshly regenerated catalyst is superior to that of catalyst that has been in contact with feed. In addition, many of the undesirable secondary reactions that occur in an FCC proceed at a slower rate than the desirable cracking reactions. Thus, short contact times favor the formation of desirable liquid products and minimize the formation of gas and coke.

The MSCC achieves extremely short contact times through the use of a radical departure in the reaction system.[6] The conventional riser is eliminated and instead, the feed is injected into a falling curtain of regenerated catalyst. The vaporized feed and catalyst move horizontally across this reaction zone and immediately enter the separation system.

A demonstration MSCC unit was built in 1990 at Coastal's Derby refinery in El Dorado, Kansas. This was done by modifying

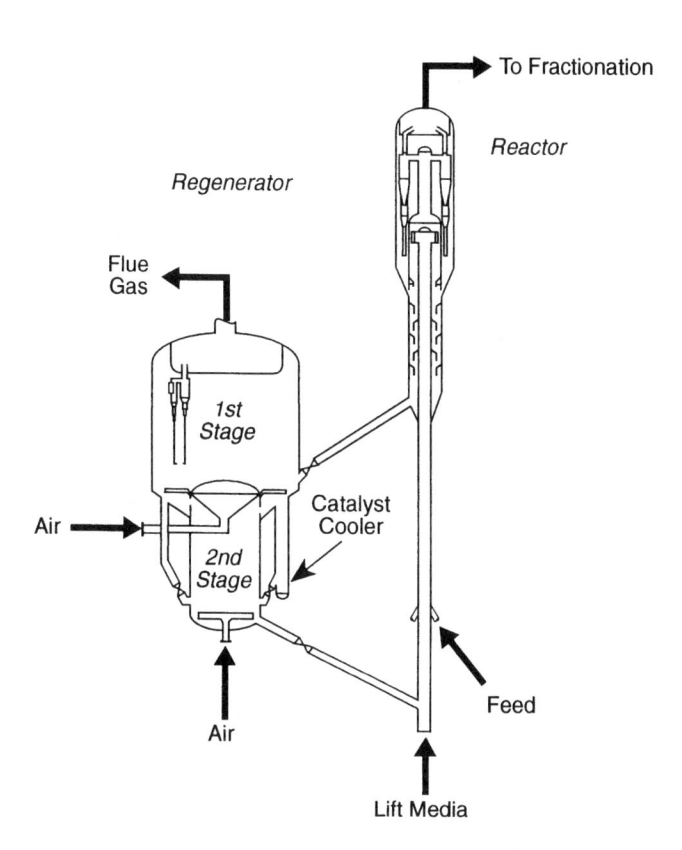

FIGURE 1-23. RCC unit. (Used with permission from UOP, RCC is a service mark of Ashland Oil Company.)

the existing 10,000 BPSD FCCU.[6] The modified unit was capable of operating in both FCC and MSCC modes. Comparative yields between these two units are given in Table 1-1.[6]

Following the successful demonstration of the process, Coastal Eagle Point Oil Company modified their 50,000 BPSD FCC to incorporate MSCC technology.

The primary advantages of the MSCC unit are the yield improvements due to short contact time cracking. In addition, the process is less costly than the traditional riser cracking system.[6]

The UOP X-design configuration uses a fluidized mixing vessel between the reactor and the regenerator.[7] Spent catalyst from the stripper and regenerated catalyst from the regenerator both flow into this vessel. The mixed catalyst flows to both the riser and the regenerator.

The blend of catalyst from the mixing vessel is cooler than the regenerated catalyst. This reduces the formation of localized high temperature zones at the feed injection point and thus, reduces thermal cracking reactions in this part of the riser.

The lower overall catalyst temperature also increases the catalyst circulation rate at any given riser outlet temperature and this further enhances catalytic reactions over thermal reactions.[7]

The spent catalyst fed to the mixing vessel still contains considerable activity. Tests by UOP on spent catalyst from several units indicates that the activity is on the order of 10 numbers lower than the equilibrium catalyst activity.[7]

The X-design layout also increases the net delta coke in the regenerator and thus the regenerator temperature. Many units operating on hydrotreated feeds with modern feed injection systems and riser termination devices actually experience low regenerator temperatures due to low delta coke.[7] For these units, the X-design will provide better regenerator operation.

Stone & Webster offer deep catalytic cracking (DCC). This process was developed in China and is intended to produce high yields of light olefins. The DCC process uses a proprietary catalyst and selected operating conditions to achieve high conversions with maximum selectivity to propylene and butylene. The reactor/regenerator section is similar to that used in an FCC. Thus it

TABLE 1-1. MSCC demonstration unit yields.

Mode Product	FCC Volume %	MSCC Volume %
Dry Gas(FOE)	7.4	3.6
LPG	21.5	20.4
Gasoline (C_5–430°F)	50.4	57.0
LCO (430–680°F)	21.6	20.6
Bottoms	9.1	9.0
Coke, Wt.%	5.8	5.5

Source: Kauff, D.A. et al. "Successful Application of the MSCC Process," 1996 NPRA Annual Meeting.

TABLE 1-2. Deep catalytic cracking yields.

Mode Products	Type I Weight %	Type II Weight %
Ethylene	6.1	2.3
Propylene	20.5	14.3
Butylenes	14.3	14.6
Amylenes	—	9.8

Source: "Refining 1996," *Hydrocarbon Processing*, November (1996), pg. 108.

is possible to switch operation from conventional catalytic cracking to deep catalytic cracking. Typical light olefin yields are given in Table 1-2.[8]

Deep catalytic cracking extends the FCC application beyond its traditional role as a primary fuels producer and into the realm of petrochemical feed stock preparation.

Future Trends

The FCC's position as the primary conversion unit in petroleum refining has been challenged from time to time by other processes. For at least the last twenty years, predictions of the demise of fluid catalytic cracking have come and gone.

The most recent challenge to the technology is the worldwide move toward environmental regulation and the drive to produce cleaner burning gasolines. Most of these programs include various controls on the make-up of both automotive gasoline and transportation diesel. Typical items of concern are:

Gasoline
    sulfur
    benzene
    olefins
    aromatics
    oxygenates

Diesel
sulfur
aromatics

Since the FCC process is a major producer of gasoline it has a significant impact on this fuel. Unfortunately, FCC gasoline contains considerable sulfur and olefins as well as some aromatics.

FCC gasoline is, in fact, the primary source of sulfur in the refinery pool and in many cases is also the major source of olefins. Both of these are considered to be undesirable.

The FCC process is also, however, the primary source for light olefins in the refinery and these compounds can be used to produce both alkylate and MTBE. Both of these products are desirable gasoline blending components which supply high octane values and do not contain any undesirable compounds.

LCO from the FCC is also high in sulfur and is also high in aromatics. This makes it increasingly less desirable as a diesel blend stock. The properties of the LCO can be improved considerably by hydrotreating, but there are limits to the aromatics saturation possible in conventional diesel hydrotreaters.

The future trends in FCC technology and operations then probably will be toward reducing the volume of FCC naphtha and increasing the yield of light olefins. Ultimately, the FCC may become less of a fuels producer and more of a feed preparation unit for downstream processes. As always, many of these advances will be lead by catalyst developments.

References

[1]*Guide To Fluid Catalytic Cracking—Part One,* W.R. Grace and Co. (1993).

[2]Blazeck, J.J., "Catalytic Cracking—Part One, History and Fundamentals," *The Davison Chemical Guide to Catalytic Cracking.*

[3]Magee, J.S., "A Guide to Davison Cracking Catalysts," *The Davison Guide to Catalytic Cracking.*

[4]Montgomery, J.A., "Catalytic Cracking—Part Two, The Evolution of the Fluid Catalytic Cracking Unit," *The Davison Chemical Guide to Catalytic Cracking.*

[5]Tippee, B., and Beck, R.J.,"Oil Demand Continues to Grow in the U.S." *Oil and Gas Journal,* July 31 (1995), p. 47.

[6]Kauff, D.A., Bartholic, D.B., Steves, C.A., and Keim, M.,R., "Successful Application of the MSCC Process," NPRA Annual Meeting (1996).

[7]Knapik, P.G., Harris, J., Hemler, C.L., Kauff, D.A., and Schnaith, M.W., "New FCC Technology," 1995 UOP Refinery Technology Conference (1995).

[8]"Refining 1996," *Hydrocarbon Processing*, November (1996), p. 108.

2
Fundamentals

Reactions

Catalytic cracking begins with the formation of a carbenium ion (R – CH_2+). Carbenium ions can be formed either by the removal of a hydride ion (H –) from a paraffin or by the addition of a proton (H+) to an olefin. These ions are formed by reactions between hydrocarbon molecules and acidic sites on the catalyst.[1]

$$R - CH_2 - CH_2 - CH_3 \rightarrow R - CH_2 - CH_2 - CH_2^+$$
(electron removal by Lewis site)

$$R - CH = CH - CH_3 \rightarrow R - CH^+ - CH_2 - CH_3$$
(hydrogen addition by Bronsted site)

The carbenium ions formed by these reactions can be primary, secondary, or tertiary (Figure 2-1). Tertiary ions are the most stable. Primary ions are the least stable and quickly undergo rearrangement to either tertiary or secondary form.[2] Thus, most carbenium ions involved in cracking reactions involve charges on interior carbon atoms. This is an important fact in catalytic cracking, which has a major effect on the types of products produced.

The next reaction step is beta scission of the carbenium ion. In this reaction the carbon to carbon bond β to the charged carbon is broken to form an olefin and a new carbenium ion.

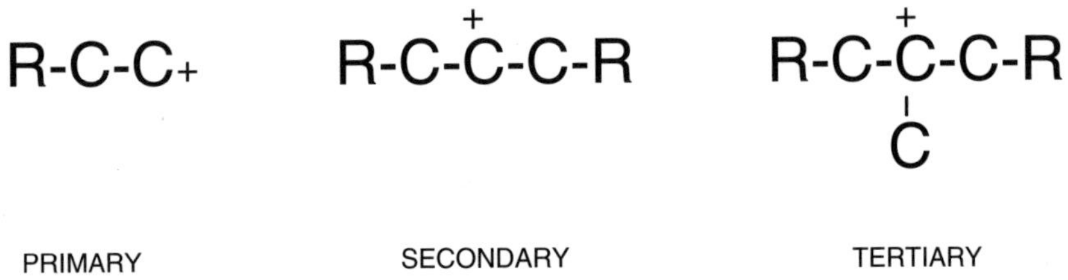

```
        +                  +                      +
  R-C-C+           R-C-C-C-R            R-C-C-C-R
                                              |
                                              C

  PRIMARY           SECONDARY              TERTIARY
```

FIGURE 2-1. Carbenium ions.

$$R - CH^+ - CH_2 - CH_2 - R \rightarrow R - CH = CH_2 + CH_2^+ - R$$
(β Scission)

Beta scission is the dominant reaction because the bond two carbons away from or β to the ionic site is the weakest carbon/carbon bond in the chain.

The fact that cracking occurs on the bond β to the charged carbon coupled with the dominance of secondary and tertiary carbenium ions means that the smallest hydrocarbon molecule formed by catalytic cracking contains three carbon atoms. This results in high yields of C_3 and C_4 LPG and the low yields of lighter gases[2].

The carbenium ion formed by β scission can undergo further cracking reactions. The olefin can also be cracked further after being converted to a carbenium ion through hydrogen addition. Thus, large hydrocarbon molecules can be cracked repeatedly producing successively smaller hydrocarbons.

As the hydrocarbon chains become smaller, however, the cracking rates become slower. Thus, as the reactions proceed the overall rate decreases.

Carbenium ions can react with paraffins through hydrogen transfer[1]:

$$CH_3 - CH^+ - CH_2 - CH_3 + R - CH_2 - CH_2 - CH3 \rightarrow$$
$$CH_3 - CH_2 - CH_2 - CH3 + R - CH^+ - CH_2 + CH_3$$

Thus, the short, less reactive ion transfers its charge to the larger, more reactive molecule further propagating the cracking reaction.

Carbenium ions are also involved in the propagation of reactions other than cracking. These include hydrogen transfer, isomerization and cyclization. These reactions have a major impact on both yield structure and product properties.

Hydrogen transfer is a bimolecular reaction that occurs on the surface of the catalyst. The hydrogen transfer reaction between a paraffin and a carbenium ion was illustrated above. Another example would be the reaction between an olefin and a cycloparaffin (naphthene). Hydrogen is transferred from the cycloparaffin to the olefin molecule. The olefin is converted to a paraffin, and the cycloparaffin becomes a cyclo-olefin. Subsequent hydrogen transfer reactions can convert the cyclo-olefin to an aromatic molecule.

Isomerization reactions occur when a carbenium ion rearranges to the tertiary form and is then involved in a hydrogen transfer reaction with a paraffin. The result is a monobranched paraffin and a carbenium ion (Figure 2-2). Since tertiary carbenium ions are the most stable, this reaction is fairly common and accounts for the large number of monobranched paraffins found in FCC products. Multibranched products are less common, but they do occur.

Another secondary reaction is cyclization of the straight chain olefins (Figure 2-3). This reaction forms a naphthene. The naphthene formed in this reaction can then be converted to an aromatic molecule through hydrogen transfer. Thus, it is possible for FCC products to contain aromatics even if the feed consists entirely of paraffinic molecules.

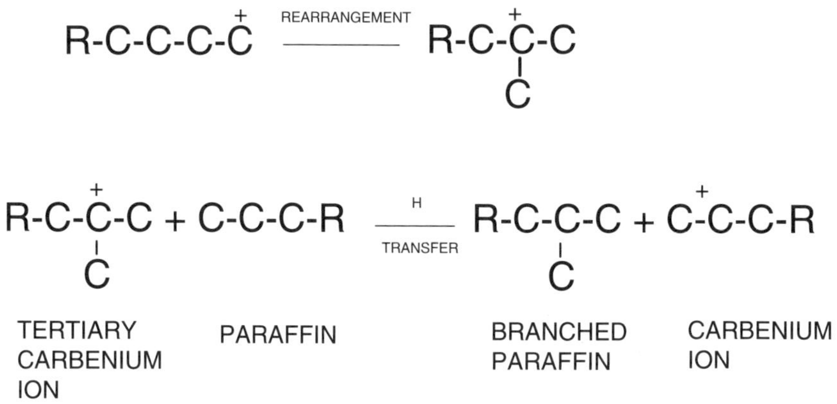

$$R\text{-}C\text{-}C\text{-}C\text{-}\overset{+}{C} \xrightarrow{\text{REARRANGEMENT}} R\text{-}C\text{-}\overset{+}{\underset{|}{C}}\text{-}C$$
$$\phantom{R\text{-}C\text{-}C\text{-}C\text{-}\overset{+}{C} \xrightarrow{\text{REARRANGEMENT}} R\text{-}C\text{-}}C$$

$$R\text{-}C\text{-}\overset{+}{\underset{|}{C}}\text{-}C + C\text{-}C\text{-}C\text{-}R \xrightarrow[\text{TRANSFER}]{H} R\text{-}C\text{-}\underset{|}{C}\text{-}C + \overset{+}{C}\text{-}C\text{-}C\text{-}R$$
$$\phantom{R\text{-}C\text{-}}C \phantom{\text{-}C + C\text{-}C\text{-}C\text{-}R \xrightarrow[\text{TRANSFER}]{H} R\text{-}C\text{-}}C$$

TERTIARY   PARAFFIN   BRANCHED   CARBENIUM
CARBENIUM   PARAFFIN   ION
ION

FIGURE 2-2.   Isomerization.

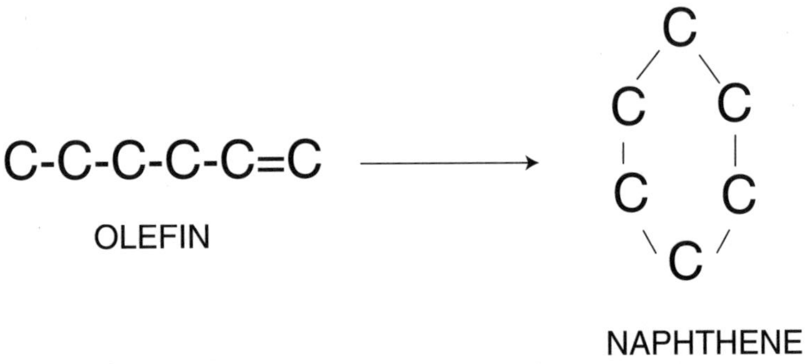

```
                            C
                          /   \
                         C     C
C-C-C-C-C=C      ⟶       |     |
   OLEFIN                C     C
                          \   /
                            C

                         NAPHTHENE
```

FIGURE 2-3. Cyclization.

Aromatic rings are not affected by primary catalytic cracking. Side chains attached to an aromatic molecule, however, can be cracked to yield olefins and shorter chain alkylaromatics. Aromatics rings can also become involved in secondary polymerization reactions. This generally leads to the formation of the multi-ring aromatic structures which are the main constituents of coke.

A summary of the various primary reactions is given in Table 2-1. Secondary reactions are given in Table 2-2.[1]

In addition to the many primary and secondary catalytic reactions taking place, thermal reactions do occur. These reactions are the result of the elevated temperatures (900°F+) in the riser.

Thermal cracking reactions begin with the formation of a free radical. Free radicals are formed by breaking a carbon/carbon or hydrogen/carbon bond in the feedstock. Since more energy is required to break carbon/hydrogen bonds, carbon/carbon separation is more common.[1]

TABLE 2-1. Primary cracking reactions.[1]

Reactants	Products
Paraffins	Olefin + Paraffin
Olefines	Olefins
Naphthene	Olefins
Alkyl Aromatic	Olefin + Alkyl Aromatic (shorter side chain)

TABLE 2-2. Secondary reactions.

Reactants	Products
Hydrogen Transfer	
Olefin + Naphtene	Paraffin + Cyclo Olefin
Carbenium Ion + Paraffin	Parafin + Carbenium Ion
Isomerization	
n-Olefins	Iso Olefins
n-Paraffins	Iso Paraffins
Cyclization	
Olefins	Naphthenes
Dehydrogenation	
Naphthenes	Cyclo Olefins, Aromatics
Aromatics	Multi-Ring Aromatics

There is little difference, however, in the energy required to break primary, secondary, or tertiary carbon/carbon bonds. Thus, the formation of methyl or ethyl free radicals is as likely as longer chains[1,3]. This explains the high yields of dry gas (C_1 and C_2 hydrocarbons) from thermal cracking reactions.

Once formed, free radicals can react further by beta scission or by hydrogen transfer with other hydrocarbons. In beta scission, the carbon bond that is β to the hydrogen deficient carbon atom breaks. The result is an olefin and a new free radical.

$$R - CH_2 - CH_2 - CH_2^* \rightarrow R - CH_2^* + CH_2 = CH_2$$

Hydrogen transfer occurs when the free radical extracts a hydrogen atom from a hydrocarbon.

$$CH_2^* + R - CH_2 - CH_2 - CH_3 \rightarrow CH_4 + R - CH_2 - CH^* - CH_3$$

The bond between hydrogen and a primary carbon atom is stronger than that between hydrogen and a secondary carbon atom. Thus, the secondary hydrogen atoms are more likely to be involved in hydrogen transfer reactions. Hydrogen atoms attached to tertiary carbon atoms are held by weaker bonds still, and if present, are even more likely to be involved in these reactions.[1]

The new free radicals formed by beta scission or hydrogen transfer are free to engage in further reactions.

Unlike carbenium ions, free radicals cannot undergo isomerization.[1] Thus, there is no increase in branching. In addition, some diolefins are formed.

Due to the poorer quality of products produced, thermal reactions are considered to be undesirable. Fortunately, these reactions generally have a slower rate than the catalytic cracking reactions.

Dehydrogenation reactions occur as a result of contaminants deposited on the catalyst. The best known contaminant is nickel, which is a strong dehydrogenation agent. Other metals which result in increased yields of hydrogen are copper and iron. In addition to increased yields of hydrogen, dehydrogenation reactions also increase the formation and condensation of aromatic molecules and thus increase the yield of coke.

Feedstock and Feedstock Characterization

The classic FCC feedstock is a heavy vacuum gas oil boiling between 650 and 1050°F. The FCC process is extremely flexible, however, and a wide range of feedstocks can be processed. In addition to the virgin gas oils, hydrotreated gas oils, cracked gas oils and deasphalted oils are common feeds to FCC units. In addition, many units process small amounts of atmospheric residue, vacuum

TABLE 2-3. FCC feeds.[1,4]

Feed	BPD	Percent
Virgin Gas Oils	2,281,363	50.9
Cracked Gas Oils	189,813	4.2
Deasphalted Oils	29,329	0.7
Hydrotreated Gas Oils	1,308,608	29.1
Atmospheric Resid	360,667	8.0
Vacuum Resid	73,343	1.6
Hydrotreated Resid	78,263	1.7
Other	172,224	3.8

TABLE 2-4. Gas oil properties.

Specific Gravity	0.89–0.93
Sulfur, Wt. %	0.30–2.00
Nitrogen, Wt. %	0.07–2.00
Conradson Carbon, Wt. %	0.00–0.20

residue, slop oils, or lube extracts. Table 2-3[1,4] lists the feedstocks processed by fluid catalytic cracking units in the U.S. during 1990.

The ability of the FCC to process this wide range of feeds into useful products has led to its use as the "garbage disposal" of the refinery. Feedstock quality can vary considerably from unit-to-unit, and from day-to-day. Even units that process "traditional" gas oil feeds can see considerable variation in feed quality. Table 2-4 contains typical properties for different gas oils.

Refineries today operate in an atmosphere of economic uncertainty. In response, many have adopted a policy of careful crude selection based on linear modeling of the refinery operations. As a result of this practice, crude slates often change on a weekly or even daily basis. Thus, the feed to the FCC can change radically over a very short time period.

Given the wide range of possible feed types and properties, proper feedstock characterization is critical to optimizing the performance of the unit. FCC feeds, regardless of their source, consist of mixtures of various types of hydrocarbon molecules. The different types present and their relative concentrations in the feed affects the yields and operation of the FCC. By identifying these hydrocarbon types and their concentrations, feed characterization allows the FCC engineer to predict the yields and operating conditions that can be expected.

Hydrocarbon Types in the FCC

Hydrocarbons can be broadly broken into four main classifications: paraffins, olefins, naphthenes, and aromatics. Each of these types represents a specific molecular arrangement. When processed in an FCCU, each of these types will produce a different slate of products.[5] When mixtures of these types (i.e., typical FCC feeds) are processed, the net products produced and their properties will be determined by the composition of the mixture.

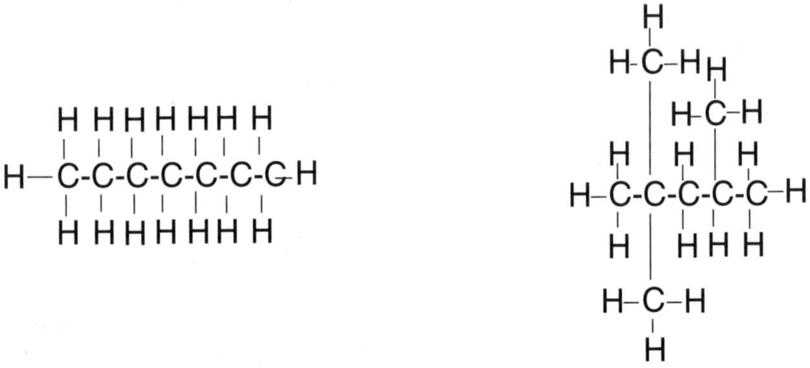

```
H H H H H H H
| | | | | | |
H—C-C-C-C-C-C-C—H
| | | | | | |
H H H H H H H
```

```
           H
           |
         H-C-H
           |  H
         H-C-H
    H    H |  H
    |    | |  |
  H-C-C-C-C-C-H
    |    | |  |
    H    H H  H
         H-C-H
           |
           H
```

NORMAL HEXANE   ISOOCTANE

FIGURE 2-4. Paraffins.

Paraffins. Paraffins are straight or branched chain hydrocarbons with no double bonds between carbon atoms. The generic chemical formula for these compounds is C_nH_{n+2}. Typical paraffins are shown in Figure 2-4. Paraffins crack easily and produce high yields of liquid products with low yields of gas and coke. In this respect, they are desirable components in FCC feeds. The gasoline produced from paraffins, however, is relatively low in octane.[5]

Olefins. Olefins are compounds that contain one or more double bonds between carbon atoms. The generic chemical formula for an olefin is C_nH_{2n}. Olefins are not present in unprocessed or virgin FCC feeds or in feeds that have been hydroprocessed. Cracked feeds, especially thermally cracked feeds such as coker gas oils or visbreaker gas oils, contain significant amounts of olefins. Olefins are also produced by primary catalytic cracking reactions. Typical olefins are shown in Figure 2-5.

Olefins are easily cracked but are also subject to most of the secondary reactions that take place in an FCC. Thus, they tend to produce higher yields of gas and coke as well as increased yields of heavier products.[5]

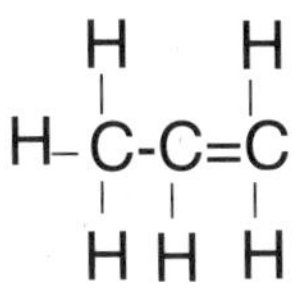

```
    H   H                          H      H H H
    |   |                          |      | | |
H – C – C = C                  H – C –C = C–C–C–C – H
    |   |   |                      |   |   |  | |  |
    H   H   H                      H   H   H  H H  H
```
PROPYLENE 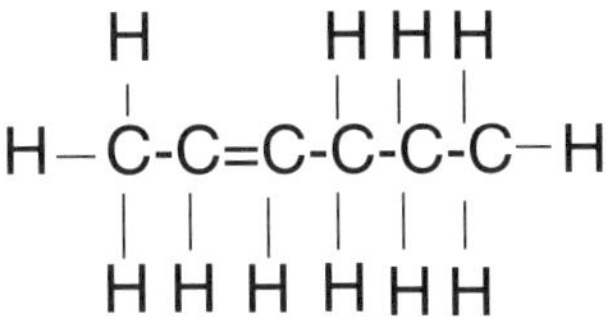   2 HEXENE

FIGURE 2-5. Olefins.

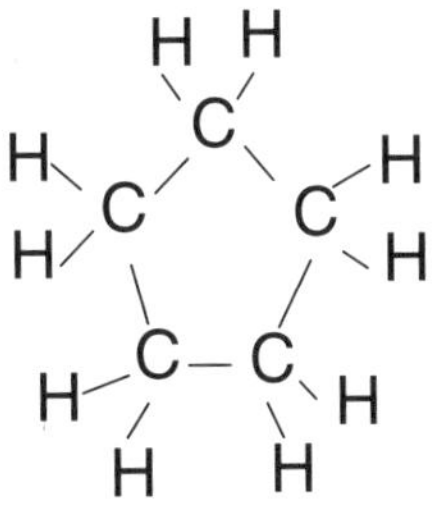

CYCLOPENTANE 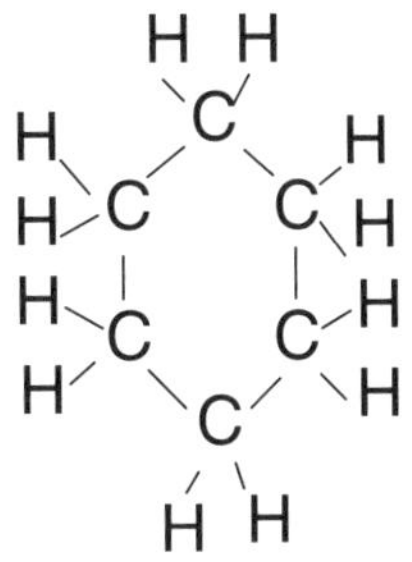   CYCLOHEXANE

FIGURE 2-6. Naphthenes.

Naphthenes. Naphthenes are saturated ring compounds with the generic formula C_nH_{2n} (Figure 2-6). Naphthenes are frequently also referred to as cycloparaffins. In addition to cracking, naphthenes can be converted to aromatics by hydrogen transfer with olefins. Many of these aromatic molecules will be single ring compounds that boil in the gasoline range. Thus, the gasoline produced from naphthenic feedstocks generally has a higher octane than that produced from paraffinic feeds.[5] Some of these aromatic molecules will, however, condense and form heavy oils or coke. Thus, the liquid yield from naphthenic feeds is generally lower.

Aromatics. Aromatic hydrocarbons are compounds containing a Naphthenes six carbon ring with three of the six carbon-carbon bonds being unsaturated (Figure 2-7). The aromatic ring

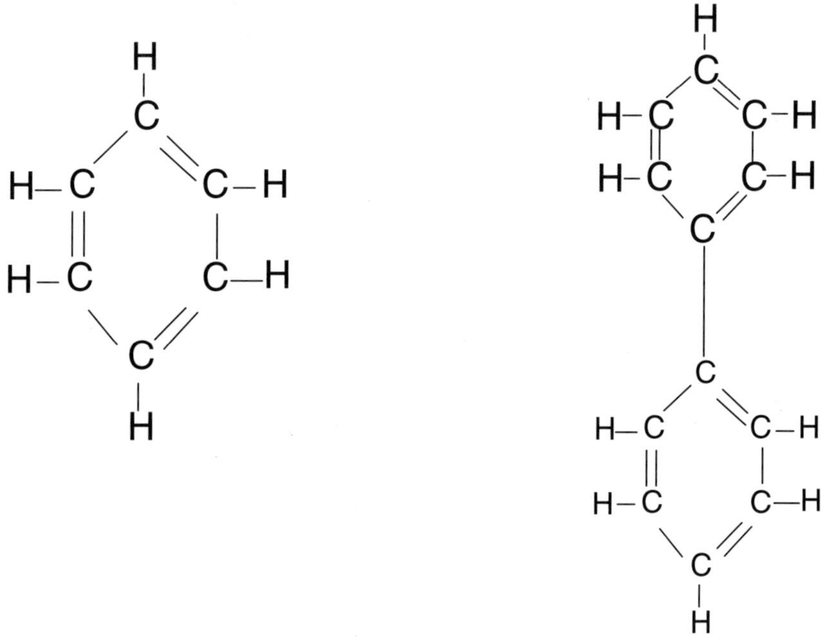

FIGURE 2-7. Aromatic compounds.

structure is very stable and these rings do not crack. Aromatic rings can, however, combine with olefins in secondary reactions. These combined compounds can eventually condense into multi-ring (polynuclear) aromatics that deposit on the catalyst as coke.

The low conversion and high yields of coke from aromatic molecules makes them highly undesirable in FCC feeds. Feeds containing a high percentage of multi-ring aromatic molecules are especially undesirable as these compounds will generally go directly to coke.[5]

Hybrid Molecules. Naphthene and aromatic compounds often contain paraffinic side chains (Figure 2-8). These side chains are subject to the same reactions as paraffinic molecules. Thus, an alkylaromatic molecule, while classified as an aromatic compound, can undergo cracking reactions on the alkyl side chain. The result of these reactions would be an olefin and an alkyl aromatic with a reduced side chain.

Similarly, an alkyl cycloparaffin could undergo both cracking of the naphthene ring or of the alkyl side chain. This molecule

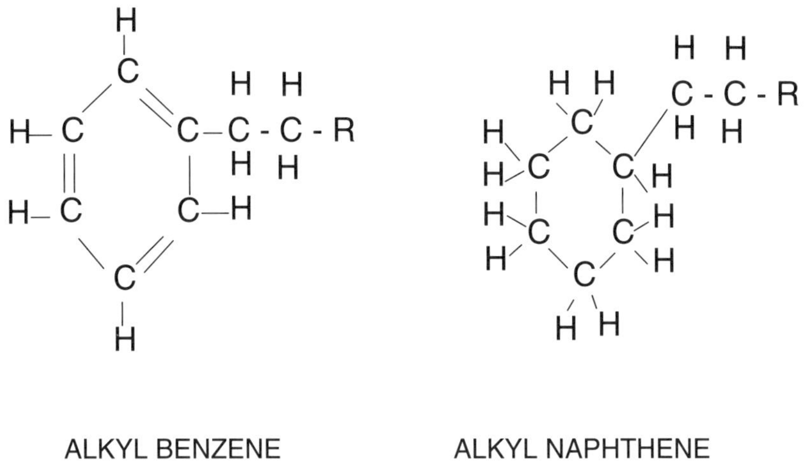

ALKYL BENZENE   ALKYL NAPHTHENE

FIGURE 2-8. Hybrid molecules.

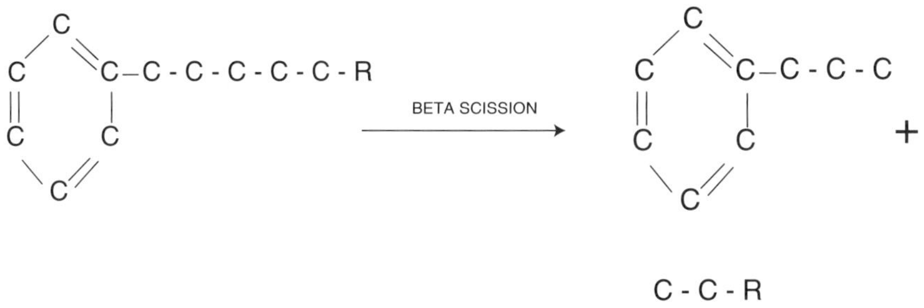

FIGURE 2-9. Hybrid molecules side chain cracking.

could also be converted to an alkyl aromatic by hydrogen transfer involving the naphthenic ring. Figure 2-9 illustrates some of the possible reactions of these hybrid molecules.

Methods for Feedstock Characterization

Early methods for evaluating FCC feedstocks relied on estimates of the relative concentration of paraffinic, olefinic, naphthenic, and aromatic molecules in the feed (PONA analysis). These methods were able to produce general correlations between FCC performance and feed quality. Their use was generally limited, however, to feeds that were similar to those used in developing the original correlations.

This limitation resulted from the inability of hydrocarbon type analysis to distinguish between ring compounds (aromatics and naphthenes) and ring compounds linked with paraffinic side changes. As discussed above, however, these two types of compounds react differently in an FCCU. This shortcoming was addressed by turning to carbon distribution as the primary method for FCC feedstock analysis.

Carbon distribution classifies feedstocks not by the types of hydrocarbon molecules in the feed but by the percentage of carbon atoms in the various molecular structures. Specifically, these methods are used to determine the values of Ca (carbon atoms in aromatic rings), Cn (carbon atoms in naphthenic rings) and Cp (carbon atoms in paraffinic chains). Cp includes the carbon atoms in paraffinic side chains linked to either aromatic or naphthenic ring structures.

The most accurate methods for evaluating the carbon distribution in FCC feeds are mass spectroscopy (MS) and nuclear magnetic resonance spectroscopy (NMR). These methods measure molecular types based on either their mass (MS) or the energy produced by hydrogen nuclei vibrating in a strong magnetic field their (NMR).

These sophisticated methods require the use of expensive equipment operated by highly trained personnel. This makes them impractical for routine refinery analysis of FCC feedstocks. Empirical correlations exist, however, that relate carbon atom distribution to easily measured physical and chemical properties.

Most empirical feed quality correlations use one or more of the following physical inspections:

specific gravity (or API gravity)
refractive index
average boiling point
aniline point
viscosity

The most well-known procedure for estimating carbon atom distribution is the n-d-M method (ASTM D-3238-85). This procedure uses the refractive index (n), the density (d), and the molecular weight (M) to estimate Ca, Cn, and Cp. Typically, the refractive index and the density are measured, but the molecular weight must be estimated. Molecular weights can be estimated using ASTM D-2502 which correlates molecular weight with viscosity. Molecular weights can also be estimated using API procedure 2B4.1 or API Figure 2B6.1,[6] or by using the procedure published by Dhulesia.[7] For feeds where the refractive index is difficult or impossible to obtain, Dhulesia also provides formulas for estimating this parameter.

The n-d-M method is widely used, but is known to be sensitive to inaccuracies in the measurement or estimation of the refractive index.

Specific gravity alone gives a general sense for the aromatic carbon content (Ca) of the feed. Higher specific gravity feeds are generally more aromatic. Since specific gravity measurements are easy to perform they can be taken daily (or more often) on FCC feeds. This provides a quick estimate of feed quality that can be used to evaluate unit behaviour on a daily basis.

Yields and Products

The primary products from catalytic cracking are olefinic LPG, cracked naphtha, and light cycle oil. By-products are tail gas, fractionator bottoms, and coke. Typical yields are shown in Table 2-5.

Volume percent conversion in catalytic cracking is defined as 100% minus the percent yields of liquid products boiling above the naphtha end point. Typically, these "unconverted" products

TABLE 2-5. Typical FCC yields.

Dry Gas, Wt. %	3.65
LPG, Vol %	28.04
Gasoline (C_5–430°F), Vol. %	57.90
Light Cycle Oil, Vol. %	18.41
Bottoms, Vol. %	6.59
Coke, Wt. %	5.18

are light cycle oil and fractionator bottoms. Weight percent conversion is defined in the same manner using the weight percent of "unconverted" products.

The volume of liquid products from a cat cracker is usually greater than the feed volume. Thus, the volume percent yield of naphtha and lighter products is greater than the volume percent conversion.

Cat cracked naphtha is a suitable blend stock for motor gasoline. The heavier fractions of this product (heavy cat naphtha) can also be used as a fuel oil cutter stock.

Light cycle oil can be used directly as a cutter for heavy fuel oil. It can also be further processed by hydrotreating into diesel blend stock.

The LPG produced in a cat cracker consists of olefins and paraffins. As more olefins than paraffins are produced in the primary cracking reactions, olefins dominate these light products. Typically, the C_3 fraction of the LPG is 65–70% propylene. Butylenes normally account for 55–60% of the C_4 fraction.

Due to its high percentage of olefins, the LPG produced in catalytic cracking is often used as a feedstock for downstream processes. Within the refinery, C_3 & C_4 olefins may be converted to motor gasoline blend stocks by several processes. The catalytic polymerization and Dimersol G processes both convert C_3 olefins into longer chain olefins which can be added to the gasoline pool. Catalytic polymerization can also process butylenes into gasoline range material.

Alkylation reacts the propylene and butylene streams with isobutane (also produced in catalytic cracking) to form isooctane and other multibranched paraffin molecules. The product from this process, alkylate, is an excellent high octane gasoline blend stock with high research and motor octane numbers as well as zero levels of olefins and aromatics.

The isobutylene produce in the FCCU can be combined with methanol to produce methyl tertiary butyl ether (MTBE). Besides providing an additional high octane blend stock, MTBE is the oxygenate of choice where regulations mandate a minimum oxygen content in motor gasoline.

In addition to these refinery uses, the light olefins produced in the FCCU can be used as petrochemical feedstocks. This is especially true of the propylene product. In situations where the refinery is located near to a polypropylene plant, the propylene may be the most valuable product produced in the FCCU.

Fractionator bottoms is most frequently used as a heavy fuel oil blend stock. In this service, it has added value as it often has lower viscosity than the virgin heavy fuel oil. In addition, its highly aromatic nature helps to hold asphaltenes in solution and thus increases the stability of the fuel oil blend.

An alternate use for fractionator bottoms is carbon black oil. To be suitable for this use, the catalyst content must be low and the bottoms product must meet certain other specifications. These specifications vary from market to market and variations in quality can also affect the price.

The tail gas from the gas concentration unit consists primarily of hydrogen, methane, ethane, and ethylene and is generally used as refinery fuel gas. If the FCCU is processing high sulfur feeds, there will be significant amounts of hydrogen sulfide in the tail gas. This is usually removed by amine treating before the tail gas is sent to fuel.

The coke produced in the catalytic cracking reaction does not leave the reactor as a product. Instead, this material deposits on the circulating catalyst and is carried from the reactor to the regenerator. The combustion of the coke in the regenerator provides the heat energy necessary for the catalytic cracking process. Essentially, the coke produced is burned internally as fuel.

Kinetics

Modern FCC simulation models use sophisticated kinetic equations to predict conversion and yields as a function of operating conditions, feed properties, and catalyst properties. In many cases, however, the effect of small changes in operation can be estimated using past operating data and simple equations. These procedures are approximations only and should be used only as a supplement to more sophisticated models.

Conversion

Cracking reactions are first order for pure components. When the feed consists of a mixture of hydrocarbons, however, the easiest to crack species react first. Thus, the feed becomes progressively harder to crack as the reaction proceeds. This results in what appears to be a second order reaction. Thus, second order kinetics are used to describe cracking reactions. The second order conversion function is defined as:

$$\varepsilon = \frac{X}{100 - X}$$

where X equals the weight percent conversion.

The second order conversion is a function of feedstock properties, catalyst type and operating conditions (after Weekman and Nace[8]):

$$\varepsilon = \frac{MATC}{100 - MATC} \frac{A \exp[-\Delta E/RT]}{(1-n)S} t_c^{-n}$$

where:

$MATC$ = effective catalyst activity (see Chapter 9)
A = function of feedstock quality
ΔE = activation energy, 18,000 BTU/lb.
R = gas constant, 1.987 BTU/lb.mole–°R
T = average reaction temperature, °R
n = catalyst decay constant
t_c = catalyst residence time, hours
S = weight hourly space velocity, hours^{-1}

The catalyst decay constant represents the effects of catalyst deactivation from coke deposition in the riser. Both A and n are functions of feed quality. The weight hourly space velocity, S, can be calculated from the cat to oil ratio:[8]

$$S = \frac{1}{(C/O) t_c}$$

Thus, the second order conversion is essentially linear with catalyst to oil ratio and varies exponentially with the reactor temperature.

Gasoline Yield

The yield of gasoline from catalytic cracking is not linear with conversion. At lower conversions, the gasoline yield increases as conversion increases. At some point, however, the

Fundamentals **63**

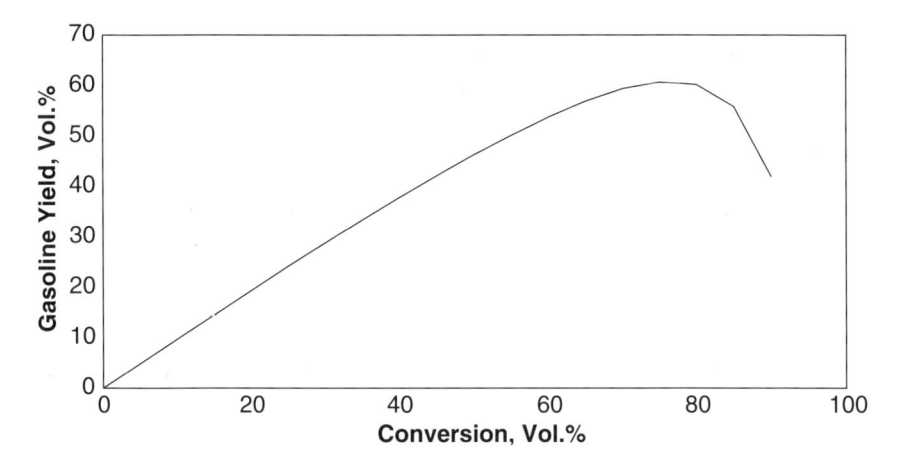

FIGURE 2-10. Gasoline yield.

yield of gasoline goes through a maximum and begins to decline as conversion increases (Figure 2-10). The conversion at which this maximum gasoline point occurs is primarily a function of the feed quality and reaction temperature. The maximum gasoline yield is a function of feed quality, catalyst type, and reactor temperature.

The maximum possible gasoline yield is inversely proportional to feed Ca and to reactor temperature. In addition, different catalyst types will give different gasoline yields from the same feed. These catalyst effects are discussed in Chapter 9.

For a given catalyst, feed and reactor temperature, the effect of conversion on gasoline yield can be estimated based on the following formula:

$$\frac{Y}{X} = B\varepsilon$$

where:

Y = weight percent gasoline yield
B = constant

The constant B is a function of feed properties, catalyst type, and reactor temperature and is always negative. Since this equation is linear with second order conversion, only two data points need to be established to determine the effect of conversion on gasoline yield for any feed and operation.

Coke Yield

For a given feedstock and catalyst, the unit coke yield is related to the second order conversion constant by the following equation:

$$\text{Coke (wt. \% fresh feed)} = k\varepsilon + C$$

where $k\varepsilon$ and C are constants. For clean feeds, C is essentially zero and the coke yield line passes through the origin. For feeds with high carbon residue, this constant equals the percentage of the carbon residue that goes to coke. This differs with feed type, but is often equal to 50 or 60 percent.

The slope of the line (k) is often referred to as the specific coke. For clean feeds, this value is equal to the coke yield divided by the second order conversion constant.

This relationship can be used to track the effect of coke yield on conversion as long as the catalyst residence time does not change significantly. If the catalyst residence time does change (for example, if the feed rate changes significantly), this will affect the coke yield.

Voorhies[9] proposed the following equation for coke yield as a function of catalyst residence time:

$$K = k\, t_c^n$$

where:

K = coke deposited on catalyst, wt.%
k = function of feed quality

The coke yield as a weight percent of fresh feed would then be:

$$\text{Coke} = K(C/O) = k(C/O)\, t_c^n$$

but,

$$(C/O) = \frac{1}{(WHSV)\, t_c}$$

therefore,

$$Coke = \frac{k\, t_c^{(n+1)}}{S}$$

which is the same form as the equation for second order conversion.

LPG

The yield of LPG plus gasoline for a given feed is approximately linear with conversion and passes through the origin. Thus, a single point is sufficient to define this line. Since the gasoline yield can be estimated based on the equation given above, the total yield of LPG can be calculated by difference.

Light Cycle Oil

The light cycle oil as a fraction of the total cycle oils is also approximately linear with conversion for any given feed. This line does not, however, pass through the origin, but instead intersects the Y axis at the percentage of LCO range material in the feed.

C_2 and Lighter

The yield of C_2 and lighter hydrocarbons is related to average reaction temperature. Thus, it is affected by the riser outlet temperature, regenerated catalyst temperature, and the feed preheat temperature. Since these light hydrocarbons are not produced by catalytic cracking reactions, there is not a simple relationship between their yield and conversion.

The gas yield is affected by both temperature and residence time. Gas yield is also affected by the catalyst type and by the presence of metals on the equilibrium catalyst. Increasing the reactor temperature will increase the yield of dry gas. Dry gas yields will also increase as the metals level on the equilibrium catalyst increases and as the hydrocarbon residence time in the reactor and transfer line increases.

Product Properties

Products produced by fluid catalytic cracking make up a significant portion of the products from the refinery. This makes the FCC product properties an important consideration in product blending. This is especially true with regard to the refinery gasoline pool. Product sulfurs and gasoline octane are critical quality parameters that are affected by the FCC.

Sulfur Distribution

Recently, the sulfur content of FCC products has received considerable attention. This is especially true with regard to the sulfur content of FCC produced naphthas. Much of this recent attention is the result of U.S. environmental regulation which requires the use of cleaner burning gasoline in large metropolitan areas.

Sulfur in gasoline has been identified as a major contributor to automobile tailpipe emissions, especially in vehicles fitted with exhaust system catalytic converters. Sulfur in the gasoline pool comes largely from FCC naphtha. For this reason, FCC naphtha sulfur content is an important variable in gasoline blending.

While most of the recent attention has been focused on FCC naphtha, the sulfur content of other cat cracker products is also of interest. Light cycle oil may be used in either diesel fuel or fuel oil blending and the fractionator bottoms will generally be sent to the fuel oil pool. The sulfur content of the coke burned in the regenerator affects the overall emissions from the refinery

The allowable sulfur levels in the diesel oil and heavy fuel oil are under increasing downward pressure. This is especially true in many countries outside of the U.S. where these products make up a significant fraction of the total petroleum demand.

These concerns make an understanding of FCC product distribution important to refinery technical service, operations, and planning personnel. Historical data are the best source for this information. When new feeds are processed, however, historical data must be reviewed and an estimated sulfur distribution must be determined based on the expected feed properties and unit operations.

Early studies on sulfur distribution in FCC products was done by Sterba.[10] In this work, pilot plant produced products were analyzed for sulfur. The tests were done using the catalysts available at the time. These included clay based catalysts as well as synthetic silica alumina and silica magnesia catalysts.

Sterba found that approximately half of the sulfur in the feed was converted to hydrogen sulfide. The sulfur content of the gasoline was found to be a function of feed sulfur, conversion, and reactor temperature.

In general, gasoline sulfur showed a strong correlation with feed sulfur. For 60% conversion and 900°F reactor temperatures, the gasoline (400°F ASTM end-point) sulfur content for most feeds fell about the line:

$$\ln S_g = 0.808 \ln S_f - 2.05$$

where:

S_g = wt.% sulfur in gasoline
S_f = wt.% sulfur in feed

Gasoline sulfur content increased with increasing reactor temperature, but decreased with increasing conversion. Sterba also found that a large fraction of the gasoline sulfur was contained in the 300°F+ fraction.

In 1971, further work was done by Wollanston et al.[11] This work was expanded by Huling and others.[12] Huling considered the distribution of sulfur in products from both hydrotreated and virgin gas oil feeds. Data in this paper were developed using a 0.2 BPD adiabatic riser pilot plant.

This study, which is generally considered the watershed work on FCC product sulfur distributions, showed that:

1. The product sulfur distribution from hydrotreated feeds is different from that of virgin feeds. Specifically, less of the sulfur remaining in hydrotreated feeds will be converted to H_2S and more sulfur will be present in the cycle oil and coke products.
2. There is a strong relationship between the sulfur in the fractionator bottoms and the sulfur in the coke.
3. The fraction of feed sulfur appearing in the gasoline is lower for hydrotreated feeds.

The authors concluded that the sulfur in desulfurized feeds was located in the harder to convert molecules. Thus, less of this sulfur was moved in to the lighter FCC products through cracking reactions.

The relationship between sulfur levels in the fractionator bottoms and in the coke held across both virgin and hydrotreated feeds. In fact, an increase in the LCO end-point from 650°F to 710°F did not alter the relationship significantly. Data from only one feed, a resid, failed to fall on the curve.

For the other feeds, the relationship between coke sulfur and fractionator bottoms sulfur was:

$$\ln S_{coke} = 1.286 \ln S_{FB} - 0.943$$

where:

S_{coke} = sulfur in coke, Wt.%
S_{FB} = sulfur in fractionator btms., Wt.%

In 1990, the reenactment of the U.S. Clean Air Act with its mandate for reformulated gasoline and very low sulfur diesel oil sparked new interest in FCC product sulfurs. Wormsbecher et al.[13] published research on the effects of catalyst type on FCC product sulfurs. As a part of this research, the sulfur compounds in full-range FCC gasoline were analyzed. These were found to consist of low molecular weight mercaptans, thiophene, methyl to butyl alkyl thiophenes, tetrahydrothiophene, and benzothiophene. Little difference was found in the sulfur distributions from various standard catalyst types.

In conducting these experiments, data were obtained for products from pilot plant cracking of low sulfur and high sulfur virgin gas oils at conversions between 63 and 80 wt.%.

Keyworth et al.[14] published the results of a similar study. This paper contained sulfur distributions for both hydrotreated and unhydrotreated feeds. In addition, data from the sulfur distribution in narrow range gasoline cuts were given. These data showed a distributions similar to that given by Sterba in 1949.

The data show a high degree of consistency from investigator to investigator. At similar conversion levels, all report approximately the same percentage of sulfur converted to H_2S for virgin feeds. Data from Keyworth and Huling also show similar effects on gasoline sulfur for hydrotreated feeds.

Based on these data, it is possible to develop correlations for sulfur in FCC products which can be used in the absence of historical data.

Product sulfur distributions as percentages of feed sulfur can be calculated by:

For virgin feeds:
H_2S = 0.30(CONV) + 18.0
Gasoline = 5.0
Coke = 0.1(CONV)
For hydrotreated feeds:
H_2S = 0.198(CONV) + 11.88
Gasoline = 0.3
Coke = 0.2(CONV)

where:

CONV = wt.% conversion (430°F cut point gasoline)

and the gasoline sulfur is for full range C_5 – 430°F true boiling point (TBP) cut point gasoline.

From the numbers calculated above and the expected weight percent yields of gasoline and coke, the sulfur content of the combined cycle oils can be calculated by difference. This can be further broken down using the correlation between coke sulfur and fractionator bottoms sulfur derived from Huling. When solved for fractionator bottoms sulfur, this gives:

$$\ln S_{FB} = 0.778 \ln S_{coke} + 0.733$$

This will give the sulfur in the fractionator bottoms product, and the sulfur in the LCO can then be calculated by difference. This last equation will be valid for LCO end-points between 650 and 710°F.

This procedure will give the sulfur contents of standard boiling range FCC products. In many cases, however, the FCC naphtha is split into light and heavy cuts (LCN and HCN). Figure 2-11 can be used to determine the sulfur content of the LCN. The HCN sulfur is then calculated by difference.

The sulfur content of the coke must be adjusted for feeds with high carbon residues. These feeds produce higher yields of coke at any given conversion. Thus, a larger fraction of the feed sulfur

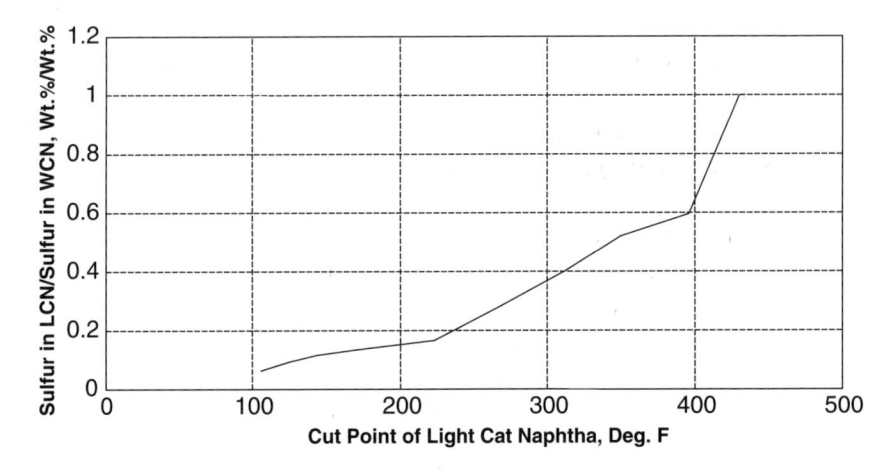

FIGURE 2-11. FCC gasoline sulfurs.

TABLE 2-6. Sulfur in coke from resid feeds.[5]

Feed	Percent of Feed Sulfur in Coke
Gas Oil	3.5
Gas Oil + 10% Resid	13.8
Gas Oil + 10% Resid	18.6

goes to the coke. Sadeghbeigi[5] published the data in Table 2-6. As these data show, the inclusion of high carbon residue stocks in the FCC feed increases the percentage of feed sulfur going to coke.

Gasoline Octane

Gasoline octanes are affected by feed type, catalyst type, conversion, and reactor operating conditions. For a given feed and catalyst, however, gasoline octane is primarily a function of reactor temperature. As a general rule, a 10°F increase in reactor temperature will give a 0.5 number increase in the research octane number (RON). The increase in motor octane (MON) will be less, typically 0.2 numbers for each 10°F increase. The RON increase is primarily due to an increase in the olefin content of the gasoline. The MON increase is primarily due to increased conversion that results from the higher reactor temperature. This increased conversion results in an increase in aromatic molecules in the gasoline product.

Feedstock effects are less pronounced. Generally feeds with higher API gravities are less aromatic and produce a lower octane gasoline. As a rule of thumb, a one number increase in API gravity will result in a 0.25 number decrease in the RON and a 0.15 decrease in the MON.[5] In a commercial unit, however, these changes may be offset by the increased conversion that normally results from changing to a lower density feed. Catalyst effects are discussed in Chapter 9.

The octane number of the gasoline can also be adjusted by changing the RVP of the gasoline. Raising the RVP increases the amount of butanes in the gasoline and this increases the gasoline octane.

Catalyst

Fluid cracking catalysts are microspheroidal particles having a particle size distribution between 10–150 microns. The primary active ingredient in catalysts is a synthetic zeolite. This component can be between 15–50 wt.% of the catalyst.

A second active ingredient found in many catalysts today is some form of alumina. This is generally referred to as active matrix and is added to improve the heavy oil or bottoms conversion activity of the catalyst. The active alumina content of FCC catalyst is generally between 0–20 wt.%.

The non-active remainder of the catalyst consists of fillers, generally kaolin clay, and binders. The fillers contribute to physical properties of the catalyst such as density, pore volume and surface area. The binders serve to hold the various constituents together in the catalyst particle.

Most catalysts available today use silica compounds as the primary binder. Other materials used as binders in certain catalysts include various aluminum compounds or clays.

In operation, fresh catalyst must be added to the unit to make up for losses and to maintain the catalyst activity. Make-up rates for units processing gas oil feeds are typically 0.15–0.20 pounds of catalyst per barrel of feed. For units processing resid feeds, the make-up rate must be higher to control the metals levels on the catalyst. For these units, the make-up rate may be as high as 1.5 pounds per barrel. When the catalyst make up rate required for activity maintenance or metals control is higher than the catalyst loss rate, equilibrium catalyst must be withdrawn from the unit to balance the fresh catalyst make-up rate.

In addition to the primary cracking catalyst, there are a number of additives which can be added to the unit catalyst inventory. Typical additives are combustion promoter (often referred to as just 'promoter') which is used to accelerate the combustion of CO to CO_2 in the regenerator; ZSM-5 which is used to increase the gasoline octane and the yield of light olefins; and SO_x adsorption additives which are used to reduce the sulfur oxides in the regenerator flue gas.

Heat Balance

The FCC heat balance determines the operation of the unit and its response to changes in independent operation variables. In a commercial unit, changes in the independent variables that affect the heat balance will result in changes in the conversion and coke yield which will bring the unit back into heat balance.

Heat inputs are:
- Heat from feed
- Heat from steam injected into the reactor/regenerator
- Heat from combustion air
- Heat produced by coke combustion

Heat outflows are:
- Heat leaving with reactor products
- Heat leaving with the regenerator flue gas
- Heat losses from the reactor/regenerator
- Heat of cracking (endothermic heat of reaction)

Essentially, the coke burned in the regenerator supplies the heat required to:

Vaporize the feed oil (fresh feed and recycle) and heat it to reactor temperature.

Supply the endothermic heat of cracking
Heat combustion air from the air blower discharge temperature to regenerator temperature.
Supply heat lost from the reactor/regenerator.
Heat coke on the catalyst from reactor to regenerator temperature.
Heat steam injected into the reactor/regenerator to exit temperature.

Independent and Dependent Heat Balance Variables

In most FCC units, the independent variables are reactor temperature, feed temperature, feed rate, and combustion air rate. Dependent variables are regenerator temperature, amount of coke produced and the heat released for each pound (or kilogram) of coke (CO_2/CO in the flue gas). To stay in heat balance, the unit must change the coke produced to adjust for any changes in heat load. This is done by adjusting the catalyst circulation rate to increase or decrease coke yield. This adjustment is automatic as the regenerated catalyst slide valve will regulate the flow of catalyst to maintain the required reactor temperature and the spent catalyst slide valve will adjust to keep the reactor bed level constant.

In a pressure balanced (Model IV) unit catalyst circulation is an independent variable (set by the pressure balance of the unit) and reactor temperature is the dependent variable. In these units, coke yield is essentially constant (unless catalyst circulation is changed manually) and the reactor temperature changes to maintain the unit heat balance.

Changes in catalyst circulation or reactor temperature have a major effect on conversion and product yields. Thus, changes in the unit heat balance will affect product production.

Coke Burned and Coke Heat of Combustion

The first step in calculating the unit heat balance is to determine the amount of coke burned and its composition. These variables are not measured directly and must be calculated from data on the air flow rate and the flue gas composition. This is done as follows:

1. From the dry air rate determine the amount of nitrogen injected into the regenerator;

$$\text{MPH } N_2 = \frac{\text{Air Rate SCFM}}{379} \times 0.791$$

2. The percentage of nitrogen in the dry flue gas can be calculated from the flue gas analysis;

$$\text{Mole\% } N_2 = 100 - \%CO_2 - \%CO - \%O_2 - \%SO_2$$

3. Since the nitrogen in the air passes through the regenerator unchanged;

$$\text{MPH Dry Flue Gas} = \frac{\text{MPH } N_2 \text{ with Air}}{\text{Flue Gas \% } N_2 \times 0.01}$$

4. Moles per hour (MPH) CO_2, CO, SO_2 and O_2 can then be calculated from the flue gas analysis and the calculated MPH of flue gas.

In addition to carbon and sulfur, the coke burned in the regenerator contains hydrogen. This is calculated by an oxygen balance around the regenerator.

5. Calculate moles of oxygen entering the regenerator with the combustion air;

$$\text{MPH } O_2 = \frac{\text{SCFH Air}}{379} \times 0.209$$

6. Oxygen consumed by elements other than hydrogen in coke;

$$\text{MPH } O_2 \text{ Consumed} = \text{MPH } CO_2 + \text{MPH } SO_2 + \text{MPH } O_2 + \frac{\text{MPH CO}}{2.0}$$

7. The difference between the oxygen calculated in (5) and (6) is the oxygen consumed to burn hydrogen in the coke. The MPH hydrogen is thus;

$$\text{MPH } H_2 = 2.0 \times (\text{MPH } O_2 + \text{With Air} - \text{MPH } O_2 \text{ Consumed})$$

8. The pounds per hour of each coke component and the total coke can now be calculated;

Lb/hr carbon = (MPH CO_2 + MPH CO) × 12
Lb/hr sulfur = MPH SO_2 × 32
Lb/hr hydrogen = MPH H_2 × 2.02
Total coke = carbon + hydrogen + sulfur

Once the amount of coke has been calculated the heat produced by its combustion can be determined. The most common method used for this calculation is to sum the heat released by the combustion of each element (carbon, hydrogen and sulfur) and to then correct this value by subtracting a "heat of desorption" to correct for the nonelemental nature of the actual coke burned.

Typical values for elemental heats of combustion and for the heat of desorption at 60°F are:[15]

Combustion
    carbon Æ carbon monoxide   3,967 BTU/Lb.
    carbon Æ carbon dioxide   14,108 BTU/Lb.
    sulfur Æ sulfur dioxide   3,978 BTU/Lb.
    hydrogen Æ water   51,485 BTU/Lb.
    Desorption (per pound of carbon)   1,450 BTU/Lb.

Regenerator Heat Balance

Once the heat released by coke combustion has been calculated, the catalyst circulation can be calculated using the regenerator heat balance. Since the heat released by coke combustion is calculated at 60°F this is the datum temperature for this calculation.

Heat inputs are:
- Heat of coke combustion (as calculated)
- Heat from air at air blower discharge temperature
- Heat from coke at reactor temperature (heat capacity equals 0.5 BTU/Lb/°F)[15]

Heat sinks are:
- Heat to raise flue gas from 60°F to flue gas temperature
- Heat losses from regenerator vessel
- Heat required to raise catalyst from reactor temperature to regenerator bed temperature

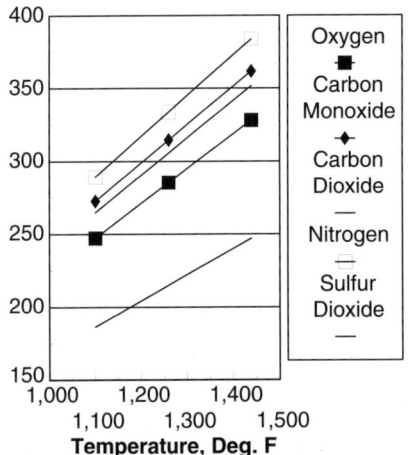

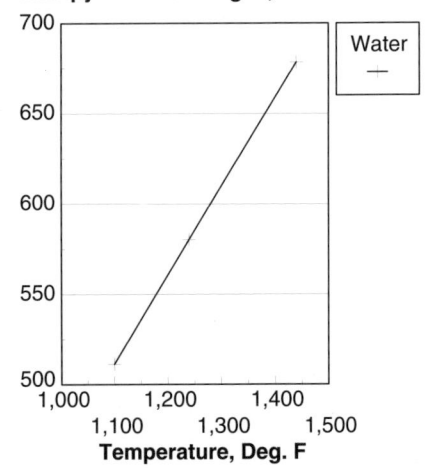

76 Chapter 2

FIGURE 2-12. Flue gas component enthalpies. (Calculated from reference 15.)

To calculate the catalyst circulation rate, the heat inputs from air and coke are calculated from the known temperatures and the flow rates. These inputs are added to the heat produced by coke combustion to give the total heat input to the regenerator.

Once the total heat in is determined, the heat leaving the regenerator with the flue gas and the radiant heat losses are calculated. The heat leaving with the flue gas is calculated based on the flue gas temperature, composition and the heat capacities of the flue gas components (Figure 2-12). The radiant heat losses can be estimated as 4.0 percent of the heat of combustion[15].

These two heat sinks are subtracted from the total heat in to give the heat to the catalyst. The catalyst circulation can then be calculated based on the temperature difference between the reactor and the regenerator. The catalyst heat capacity is 0.265 BTU/Lb/°F[15].

Reactor Heat Balance

The hot catalyst from the regenerator flows to the riser/reactor system where it provides the heat necessary to:

- Vaporize the liquid oil feed
- Raise the vapors to reactor temperature

- Raise steam injected into the riser to reactor temperature
- Supply the endothermic heat of cracking
- Supply heat lost from the riser and the reactor vessel

By convention, the heat content of the products leaving the reactor are assumed to be the same as vaporized fresh feed at reactor conditions. Thus, for the purpose of the reactor heat balance, the heat to the feed is equal to:

$$\Delta H_{feed} = H_{OV} - H_{IL}$$

where:

H_{OV} = enthalpy of feed vapors at reactor outlet conditions
H_{IL} = enthalpy of liquid feed at feed inlet temperature

Heat losses can be estimated as 2.5% of the heat transferred to the riser by the catalyst.[15] The coke heat of adsorption (equal to the heat of desorption in the regenerator) is added as a heat input.

Most reactor heat balances are closed by calculating the heat of cracking.

The heat of cracking contains the endothermic heat of reaction. It also contains the corrections required to account for the various simplifying assumptions made in the heat balance procedure. For this reason, the heat of cracking is a function of the heat balance procedure used, and values obtained using different procedures cannot be compared.

The heat of cracking is normally expressed as BTU/Lb. of fresh feed and normally falls in the range 100–250 BTU/Lb. of feed.

Mass Balance and Test Runs

Regular material balances are essential to maintaining optimum FCC performance. Material balances serve to identify the current yield structure and serve as the basis for changes in unit operation to increase product value. A routine material balance should be calculated on a daily basis. This is usually sufficient for normal operations. In some cases, a more detailed evaluation of the unit performance may be necessary. In these cases, a unit test run may be required.

Routine Material Balance

Routine material balances normally involve collecting flow rates for the feed, product, and combustion air streams. In addition to these flows, densities or compositions must be determined for each product stream; this normally involves sampling and laboratory analysis. At a minimum, it will be necessary to determine the densities of each naphtha product, the light cycle oil, and the fractionator bottoms product. The LPG and fuel gas product streams are normally sampled for component analysis. The flue gas analysis is determined by online analyzers, or by sampling and laboratory analysis.

These data are sufficient to calculate a rough material balance. To improve the overall accuracy, the measured flow rates should be corrected for the actual flowing density.

This routine material balance is usually sufficient for daily unit monitoring. The product yields developed, however, are on an as produced basis. Thus, these yields cannot be readily compared with data from other units from periods of operation with different product cut points.

Test Runs

From time to time, a more exact evaluation of the unit yields and operation may be necessary. This will require a rigorous test run. Test runs involve close monitoring of the unit for a set period of time. In many cases, this will require the commitment of additional refinery resources to FCC operations. This may result in reduced support to other units or additional costs. For these reasons, routine monitoring test runs are generally not justified.

Once a test run has been scheduled, the necessary resources should be committed. Test runs involve frequent sampling and analysis of the FCC product. Additional operations and laboratory personnel may be required to process the additional load. Typical samples and analysis schedules are given in Table 2-7. Extraneous gas and liquid streams that are fed to the FCC gas plant must also be sampled and analyzed on the same schedule as the product streams.

The instruments that will be used to monitor the unit yields and operation (Table 2-8) should be calibrated and zeroed. If the feeds to the product recovery section include gas or liquid streams

TABLE 2-7. Test run samples and analysis.

Stream	Analysis	Frequency
Fresh Feed	Gravity	Shift
	Distillation Sulfur	Start and End of Test Run
	Carbon Residue	
Dry Gas	Composition (by GC)	Shift
LPG	Composition (by GC)	Shift
Naphthas	Gravity	Shift
	Distillation	Shift
Cycle Oils	Gravity	Shift
	Distillation	Shift
Fractionator Bottoms	Gravity	Shift
	Distillation	Shift
Flue Gas	Composition	Shift

TABLE 2-8. Test run instruments.

Flows
    Fresh Feed & Recycle (if any)
    Air to Regenerator
    Dispersion Steam
    Stripping Stream
    Products
Temperatures
    Reactor
    Regenerator (Dense & Dilute)
    Air from Air Blower
    Fresh Feed
    Recycle (if any)
Pressures
    Reactor
    Regenerator

from other process units, these meters must also be zeroed and calibrated.

The unit feed quality should be as constant as possible during the test run. If feasible, the feed should come from a single tank. If tanks must be switched during the test, then each new tank should be sampled and analyzed before the switch.

Unit operation should be stabilized at least 24 hours prior to the formal beginning of the test run. At this time, a full set of samples should be analyzed and a full set of operating data should be taken. These data should then be used to make a complete heat and material balance around the unit.

If the results of this preliminary heat and material balance are within test run tolerances then the test run can go forward. If not, the test run should be delayed until the cause of the discrepancy can be found and corrected.

Once the test run begins, the unit operation should be held as steady as possible. In the event of a short-term upset, the test run should be put on hold until the unit can be brought back to test run conditions.

Test run lengths will depend on the purpose of the test run. Test runs used to evaluate unit operations for internal purposes often last only one shift. Test runs used to establish guaranteed performance often last several days.

During the test run, operating data are recorded on a regular schedule, usually several times per shift. In units with modern control systems and automated information systems, this is often done automatically.

All flow rates must be corrected for flowing density. These corrected flow rates are then averaged over the test run period. The average flows along with average densities are then used to calculate a raw material balance. If this balance is within $\pm 2\%$, the test run is usually considered valid.

The mass balance is normalized to 100 percent. This can be done by either adjusting the feed rate or by adjusting the product flow rates (prorated). This normalized material balance gives the product yields for the test run.

As with routine unit monitoring material balances, these produced yields cannot be readily compared with other units or with past operations with different product cut points. Since test runs are generally used to evaluate changes in unit performance over time, it is generally best to convert the raw yields to standard yields based on perfect fractionation. The most frequently used "standard yields" are:

- hydrogen sulfide
- hydrogen
- methane
- ethane
- ethylene
- propane
- propylene
- isobutane
- n-butane
- butylenes
- C_5 – 430°F TBP cut point naphtha
- 430–650°F light cycle oil
- 650°F+ fractionator bottoms
- coke

Conversion from "as produced" yields to these standard yields requires the following laboratory analyses:

- dry gas composition
- LPG composition
- light ends in naphtha (volume percent of all C_4 and lighter components)
- distillations for the naphtha, light cycle oil, and fractionator bottoms streams

If the normal laboratory distillations are by ASTM procedure D86 for light materials or D1160 for heavier materials, these must be converted to TBP distillation. This can be done using API Figure 3A.1.[6]

Once these data are available, the as produced yields are converted to standard yields. Pure components (C_4 and lighter) are grouped together. The volume percent yields of naphtha, LCO, and fractionator bottoms are corrected to standard cut points using the TBP distillations for each stream. All material boiling below 430°F is considered to be naphtha. Similarly, material boiling between 430 and 650°F is assigned to the LCO yield and the 650°F material is placed in the fractionator bottoms yield.

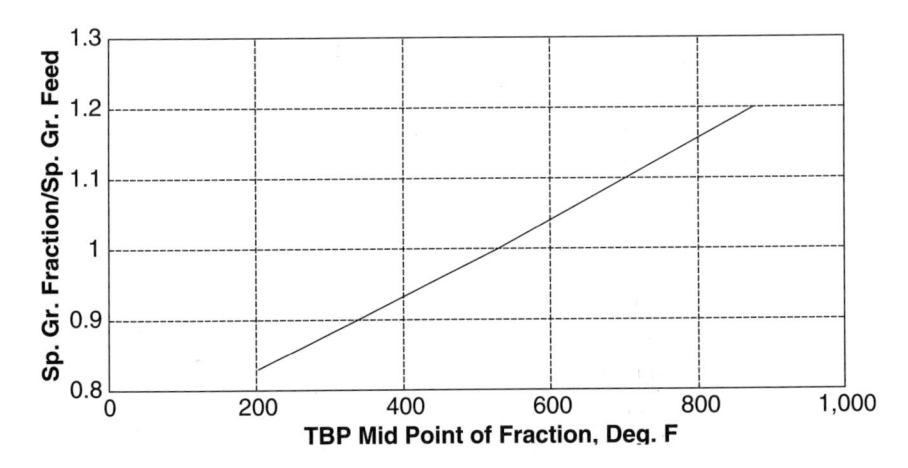

FIGURE 2-13. FCC product densities.

To correct the weight percent yields, the density of the materials transferred from one product stream to another must be known. This can be estimated by plotting a curve similar to the one shown in Figure 2-13. To plot this curve, the gravity of the gasoline, LCO and fractionator bottoms stream as a fraction of the feed gravity are plotted against their mid boiling points. The gravities of the fractions moved between products can then be estimated based on the mid points of each fraction.

Reaction Mix Sampling

Reaction mix sampling or RMS is an alternate method for determining yields. Traditional material balance procedures use product flow rates determined by stream flow meters or from product tank gauges. These rates are subject to instrument calibration error. In addition, the instrument readings must also be corrected for product density variations.

In many cases, these inaccuracies are large enough to mask small changes in unit performance (as would be expected from a catalyst or feedstock change). In addition, in many refineries, the FCC gas plant also processes extraneous gas and liquid feeds from other refinery process units. When this is the case, the measured product flow rates must also be corrected for these extraneous feeds.

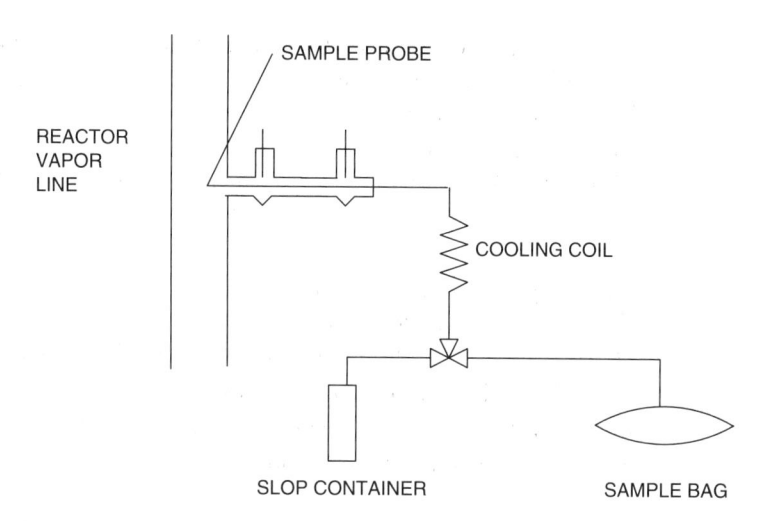

FIGURE 2-14. Reaction mix sampling. (Adapted from Reference 16.)

Reaction mix sampling eliminates these problems. In an RMS test, a sample probe is inserted into the reactor vapor transfer line (Figure 2-14).[16] Product vapors are withdrawn and cooled and collected in a sample bag. The collected vapors and liquids are then measured and analyzed to determine the product yields.

Measurement and analysis are done under controlled laboratory conditions. Thus, the accuracy is improved over that possible from a typical test run. Accuracy is, however, sensitive to sampling technique. The sampling team should be made up of skilled technicians. This makes the procedure somewhat more labor intensive than more traditional test run methods.

Reaction mix sampling data does not include the coke yield. This must still be determined based on the flue gas composition and the air rate.

Hydrogen Balance

A hydrogen balance can be used to check the accuracy of any set of yields. These yields may be from a recent unit material balance or may be yield estimates provided by catalyst vendors or FCC technology licensors.

In the first case, a hydrogen balance serves as a check on the measured flow rates and the product properties and compositions. In the case of yields provided by vendors or licensors, a hydrogen balance can be used to identify an unrealistic estimate.

The following information is needed to perform a hydrogen balance:

1. Dry gas yield and composition
2. LPG yield and composition
3. Gasoline yield, distillation, gravity, and light ends
4. LCO yield, distillation, and gravity
5. Fractionator bottoms yield, distillation, and gravity
6. Fresh feed yield, distillation, and gravity

The hydrogen contents for the C_4 and lighter products are known from their composition:

Compound	Wt.% Hydrogen
hydrogen	100.00
hydrogen sulfide	5.92
methane	24.13
ethane	20.11
ethene	14.37
propane	18.29
propylene	14.37
butanes	17.34
butylenes	14.37

The hydrogen content of the fresh feed and of the gasoline and heavier liquids can be estimated from the distillation and gravity using the following equation.[17]

$$\text{Wt.\% Hydrogen} = 0.171 \times \text{API} + 3.718 \times \ln(\text{VABP-C}) - 16.43$$

where:

API   = Gravity, Degrees API
VABP = Volumetric Average Boiling Point, °F
C   = 24 for gas oil feeds
      = 35 for resid feeds
      = 24 for naphthas
      = 10 for light cycle oil
      = 17 for fractionator bottoms

Alternately, the hydrogen content of these steams can be estimated using API Figure 2B6.1.[6]

Finally, the hydrogen content of the coke is determined in the same manner as detailed above under heat balance.

The bulk of the hydrogen in the products is in the gasoline and heavier liquids. Thus, the accuracy of the hydrogen balance is dependent on an accurate hydrogen content estimate for these streams. Since these estimates rely on empirical correlations, a closure of ±5 percent should be considered reasonable. If the closure differs by more than 5 percent, however, the yields should be considered suspect.

Pressure Balance

Fluid catalytic cracking units function by circulating the solid catalyst between the reactor and the regenerator. The key to this circulation is the pressure balance.

When properly fluidized, FCC catalyst will act as if it were a fluid. That is, it will flow under the influence of gravity or a pressure differential. Thus, the reactor/regenerator system of an FCC is essentially the same as the water-filled system shown in Figure 2-15. If the pressures in the two vessels are equal, the water levels are also equal in Figure 2-15 (a). If, however, the pressure in vessel 1 is lower than that in vessel 2, the water level will rise as

a. P1=P2
EQUAL LEVELS

b. P1<P2
UNEQUAL LEVELS

c. P1<P2+AIR INJECTION
CIRCULATION

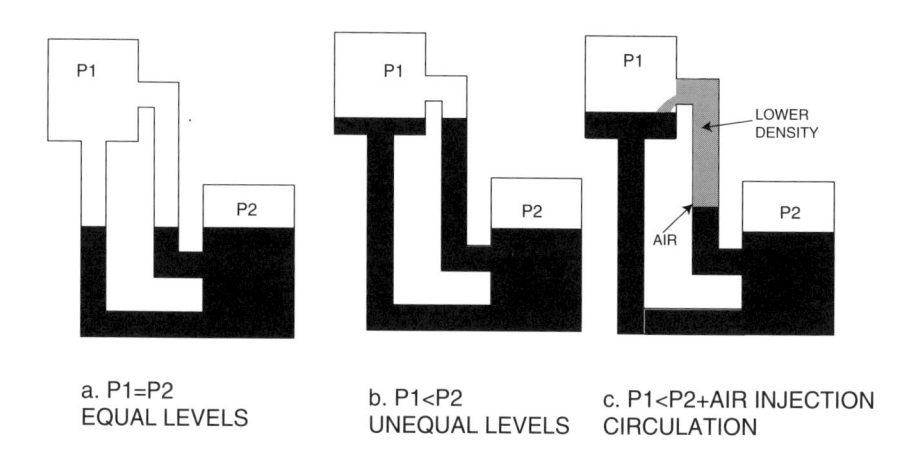

FIGURE 2-15. Water system pressure balance.

shown in Figure 2-15 (b). Finally, if air or some other gas is injected into the riser connecting the two vessels in Figure 2-15 (c), the level of water in this line will rise further due to the decrease in density. If sufficient air is injected into the riser, then the level will overflow into vessel 1 and a circulation loop will be established.

This is the process used to circulate fluidized catalyst in an FCC. Typically, the reactor vessel is equivalent to vessel 1 and the regenerator is equivalent to vessel 2. The density in the riser is low due to the presence of feed and product vapors. Thus, catalyst flows up the riser and into the reactor.

Since the catalyst in the reactor and regenerator beds is fluidized, pressure increases from the top to the bottom of the beds. This is also true in the standpipes between the reactor and the regenerator and between the regenerator and the base of the riser. This static head provides much of the pressure differential used to drive catalyst circulation. Typical catalyst densities are shown in Figure 2-16.

Since the catalyst circulation is a closed loop, the sum of the pressure increases due to static head must exactly equal the sum of the pressure losses. In most units (other than Model IVs), this

RISER
4-6

REACTOR
DILUTE PHASE 0.1-1.0
BED 35-40

STRIPPER
35-40

REGENERATOR
DILUTE PHASE 0.7-1.0
BED 25-35

STANDPIPES
35-40

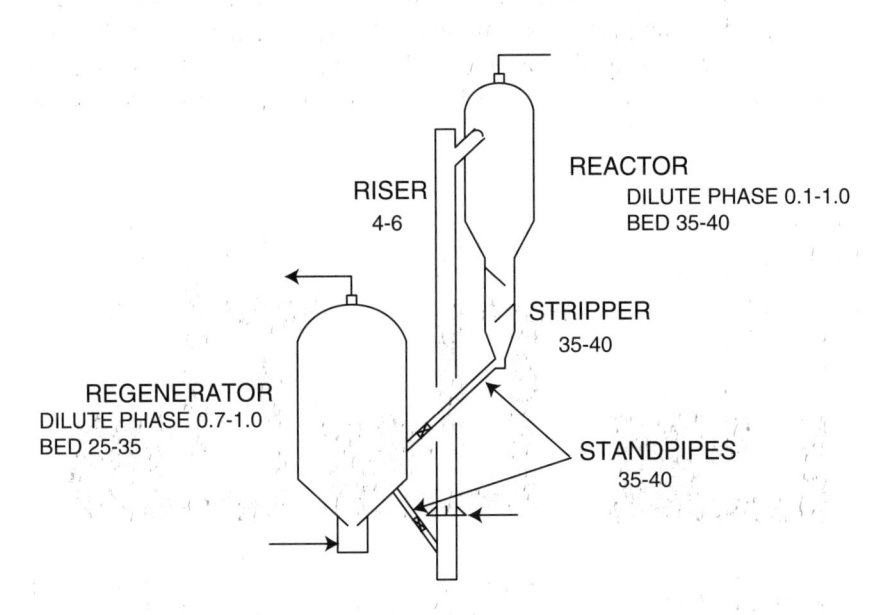

FIGURE 2-16. FCC densities (all densities in lb./ft^3.).

TABLE 2-9. Typical pressure balance.

Reactor pressure	23.0 psig
Reactor dilute phase (35 ft. @ 0.1 lb/ft^3)	+0.02 psi
Reactor bed (32 ft. @ 30 lb/ft^3)	+8.3 psi
Spent catalyst standpipe (20 ft @ 35 lb/ft^3)	+4.9 psi
Pressure above spent cat slide valve	36.222 psig
Slide valve pressure drop	−5.0 psi
Pressure at standpipe exit	31.22 psig
Regenerator bed (4 ft @ 28 lb/ft^3) above slide valve exit	−1.0 psi
Regenerator dilute phase (30 ft @ 1.0 lb/ft^3)	−0.2 psi
Regenerator pressure	30.02 psig
Regenerator dilute phase (30 ft @ 1.0 lb/ft^3)	+0.2 psi
Regenerator bed (15 ft @ 28 lb/ft^3) to regenerator cat standpipe	+2.9 psi
Regenerator cat standpipe (15 ft @ 35 lb/ft^3)	+3.6 psi
Pressure above regenerator cat slide valve	36.72 psig
Regenerator slide valve pressure drop	−6.4 psi
Pressure at base of riser	30.32
Riser pressure drop	−7.31 psi
Reactor dilute phase (15 ft @ 0.1 lb/ft^3)	0.01
Reactor pressure	23.0 psig

pressure balance determines the pressure drop available across the catalyst slide valves.

A typical pressure balance is shown in Table 2-9. In this case, the pressure drops are 5.0 and 6.4 psi for the spent and regenerated catalyst slide valves respectively. Note that the slide valve pressure drops are a function of the pressure balance only. The slide valve opening does not affect these pressure drops. Instead, the available pressure drop and the catalyst circulation set the slide valve opening.

The pressure drop across the spent catalyst slide valve is set by the reactor regenerator pressure differential and the catalyst heads in the stripper, standpipe, and regenerator. Slide valve opening has no effect on these factors, and the spent catalyst slide valve pressure drop is independent of slide valve opening.

The regenerated catalyst slide valve pressure drop is determined by the reactor and regenerator pressures, the catalyst head in the regenerator bed and the standpipe and the riser pressure drop. The riser pressure drop is a function of both feed rate and catalyst circulation. As the regenerated catalyst slide valve opens, the catalyst circulation increases. This causes an increase in the riser pressure drop and, thus, a decrease in slide valve differential. The change in differential is not, however, due to the increased open area. Instead, it is due to the increased riser pressure drop.

References

[1] *Guide to Fluid Catalytic Cracking—Part One*, W.R. Grace & Company (1993), pp. 15–17.

[2] Blazek, J.J, "Catalytic Cracking—Part One History and Fundamentals," *The Davison Chemical Guide to Catalytic Cracking,* p. 5.

[3] Hatch, L.F., *Hydrocarbon Processing,* February (1969), pp. 77–86.

[4] National Petroleum Refiners Association, *Survey of U.S. Refining Industry Capacity to Produce Reformulated Gasoline—Part A,* January (1991), p. 5.

[5] Sadeghbeigi, R., *Fluid Catalytic Cracking Handbook,* Gulf Publishing, 1995.

[6] American Petroleum Institute, *Technical Data Book—Petroleum Refining* (1983).

[7] Dhulesia, H., "New Correlations Predict FCC Feed Characterization Parameters," *Oil & Gas Journal,* January 13 (1986), pp. 51–54.

[8] Weekman, V.W. and Nace, D.M, "Kinetics of Catalytic Cracking Selectivity in Fixed, Moving and Moving Bed Reactors," *AIChE Journal,* May (1970), pp. 397–404.

[9] Voorhies, A., "Carbon Formation in Catalytic Cracking," *Ind. Eng. Chem.* (1972), p. 318.

[10] Sterba, M.J., "Sulfur Content of Catalytically Cracked Gasolines," *Industrial and Engineering Chemistry,* December (1949).

[11] Wollanston, E.G., Forsythe, W.L., and Vasalos, I.A., "Sulfur Distribution in FCC Products," *Oil and Gas Journal,* Vol. 69, No. 31 (1971).

[12] Huling, G.P., McKinney, J.D, and Readal, T.C., "Sulfur Distribution in High Conversion Riser Catalytic Cracking," 40th API Midyear Refining Meeting, May 13 (1975).

[13] Wormsbecher, R.F., Chin, D.S., Gatte, R.R., Albro, T.G., and Harding, R.H., "Catalytic Effects on the Sulfur Distribution in FCC Fuels," NPRA Annual Meeting, March (1992).

[14] Keyworth, D.A., Reid, T.A., Asim, M.Y., and Gilman, R.H., "Offsetting the Cost of Lower Sulfur in Gasoline," NPRA Annual Meeting, March (1992).

[15]"Questions Frequently Asked About Cracking Catalysts," *Catalagram,* W.R. Grace & Company, No. 57, p. 24.

[16]Hsieh, C.R. and English, A.R., "Two Sampling Techniques Accurately Evaluate Fluid-Cat-Cracking Products," *Oil and Gas Journal,* June 23 (1986), pp. 38–43.

[17]Dede, J., "Interpreting Yield Estimates," *Davison Catalagram* (1991), pp. 5–7.

Riser/Reactor Design and Operation

Catalytic cracking takes place in the vapor phase normally at temperatures between 900–1050°F. In modern riser reactors, the contact time between the reactants and the catalyst is on the order of 1.0–3.0 seconds. In this short contact time, feed must vaporize, interact with the active sites on the catalyst and undergo cracking reactions. Thus, proper design of the reaction system is crucial to achieving satisfactory results.

In most FCCUs, the actual reaction takes place in the riser. The reactor vessel itself serves primarily as a container for the cyclones used to separate the catalyst from the product vapors and as a collection vessel for the spent catalyst leaving the riser.

The riser itself can be divided into three sections: feed injection, riser/reactor, and riser termination. Each of these sections plays an important but different role in achieving the desired results.

Feed Injection Section

The purpose of the feed injection system is to insure rapid vaporization of the oil feed in the riser. Since the desired cracking reactions occur in the vapor phase, rapid vaporization is desirable. Conversely, unvaporized feed will eventually deposit on the catalyst or on the reactor walls where it will form coke.

The heat required to vaporize the oil comes from the hot catalyst entering the riser. This heat must be transferred to the oil.

The most effective means for heat transfer between the solid catalyst and the liquid oil is by direct contact of these two species. Thus, a well-designed feed injection system will maximize the potential for contact between the feed and catalyst at the bottom of the riser.

To achieve this goal, the feed injection system must atomize the liquid oil into droplets that can be distributed in the riser. These droplets must then be spread across as much of the riser cross section as possible. Finally, the catalyst entering the riser must also be distributed across the riser area. In addition, to minimize catalyst back mixing, the catalyst flow should be directed along the axis of the riser.

The feed injection system consists of two components: the nozzle(s) and the catalyst preentry zone. The nozzles atomize the oil and spread the oil droplets across the riser cross section. The catalyst preentry zone establishes the direction of the catalyst flow and determines the catalyst distribution at the feed injection nozzles. Proper design of both of these components is critical to the performance of the feed injection system.

Feed Nozzles

Most recent design changes in feed injection systems have focused on the design and function of the feed nozzles. Early FCCs generally used a simple pipe injector located at the base of the riser. In these designs, the riser was essentially a transfer line to move the catalyst to the reactor bed.

With the advent of zeolite catalyst, the function of the riser changed from a transfer line to the primary reaction zone in the FCCU. With this change, the design of the feed injection system increased in importance.

The first improvement in feed injection design was a switch from a single pipe injector to multiple nozzles located on the riser circumference. Nozzle systems of this type were used in Esso Model IV units (bed crackers) and in Kellogg Orthoflow F designs (riser crackers). The purpose of this change was to provide a more even distribution of the oil over the riser. In addition, the placement of the nozzles around the riser circumference tended to offset the tendency of the catalyst to migrate to the riser walls (this tendency is present in all dilute phase upflow solids transport).

At approximately the same time as this development, UOP introduced the "shower head" feed distributor. This improvement consisted of a multiple orifice distributor added to the older single pipe injector system. Others, mostly operating companies, have used similar configurations with multiple injection points located in the base of the riser. These systems provided improved oil distribution but did not address the tendency of the catalyst to collect on the riser wall.

The early multiple nozzle systems were essentially collections of open pipe nozzles. These systems provided little atomization of the oil. In addition, oil leaving the end of an open pipe nozzle flows as a single stream (much like water leaving the end of a garden hose). Thus, there was no dispersion of the oil across the riser cross section.

With the introduction of complete CO combustion in the regenerator, the temperature of the catalyst entering the riser increased. This led to increased thermal cracking at the base of the riser. This resulted in lost liquid yields as well as problems with product quality.

To address this problem, it was necessary to improve feed/catalyst contacting in the feed injection zone. To this end, the nozzle spray patterns were improved by modifying the tips of the feed injection nozzles. Generally, two different spray patterns were used, hollow cones and flat sprays. These spray patterns improved the distribution of the oil across the riser. Atomization was, however, only improved slightly.

In the late 1970s, Total Petroleum and UOP each developed feed nozzles to be used on residual fluid catalytic cracking units (RFCCUs). Both of these nozzle designs were based on improved atomization of the feed oil. Improved atomization would in theory improve feed vaporization due to the higher surface to volume ratio of the oil droplets. In addition, improved atomization would increase the number of oil droplets present in the feed injection zone and thus maximize the contact between oil and catalyst.

In recent years, others have also developed feed nozzles with improved atomization. Atomizing nozzle designs are available from Exxon, Kellogg, UOP, and Lummus.

Catalyst Pre-entry

Early FCC designs with the feed nozzle(s) located at the base of the riser did not include a catalyst preentry zone. The hot catalyst from the regenerator flowed down a standpipe and entered the riser at the same location as the feed.

As the catalyst entered the riser, it changed direction (from that of the standpipe to that of the riser), contacted the feed, and was accelerated to riser velocities. All of this occurred at the same point in the base of the riser.

This resulted in turbulence and back mixing in the feed injection zone. Essentially, this zone was a back mixed catalyst bed with considerable recontacting of catalyst with freshly injected oil. This initial back mixed step produced poorer yields than would be expected from a plug flow riser. In addition, this initial step was at the maximum temperature in the riser and this resulted in increased thermal cracking.

With the advent of multiple nozzles, located around the riser circumference, it became possible to preaccelerate the catalyst before it entered the feed injection zone. This was done by introducing the catalyst into the riser below the feed nozzles. Steam or gas was also introduced at this point to fluidize the catalyst and to direct the flow up the riser.

With this system, the catalyst does not change direction at the feed injection point. In addition, since the catalyst is already flowing upward, there is less acceleration required to reach riser velocities. These two facts minimize catalyst back mixing at the feed nozzles.

Depending on the velocity of the fluidizing gas, the catalyst in the preentry zone can be in any of several fluidization states. At very low velocities (below minimum bubbling), the catalyst would be in a state of particulate fluidization. Since minimum bubbling velocities for most catalysts are very low (below 0.01 ft/sec), this fluidization state does not normally exist in a commercial FCC.

Above minimum bubbling velocities up to about 2.5–3.0 ft/sec, the catalyst bed would be in either the bubbling bed or the slug flow regime. These two differ only in that in slug flow, the bubble diameter is at or near the diameter of the bed. Slug flow can exist in the catalyst preentry zone if the length to diameter ratio is sufficiently large.

Above 3.0 ft/sec and below the choke velocity, the bed would be in the turbulent/fast fluid bed regime. In these regimes discrete bubbles do not exist. In actuality, these two states represent a single continuum of fluidization and are differentiated only by the presence or absence of an observable dilute phase above the bed. In both states, the catalyst bed is still in dense phase.

Above the choke velocity (6–10 ft/sec in most risers), the catalyst is in dilute phase transport. In this regime, the catalyst is dispersed through the vapor phase. Individual particles are widely separated and the density is low.

To maximize the possibility of catalyst/oil contact, the catalyst entering the feed injection zone should be in the turbulent/fast fluid bed regime. This will assure high catalyst density without the possibility of catalyst slug flow. Superficial gas velocities in the catalyst preentry zone should thus be in the 3.0–10.0 ft/sec range (the actual value will be a function of the catalyst mass flow rate and the diameter of the preentry zone). In addition, the vertical length of the preentry zone must allow for the catalyst flow to become fully developed across the riser cross section. This normally requires a distance of 6.0–10.0 feet.

Preferred Feed Injection System

The preceding discussion points to a preferred feed injection system. The best performance can be achieved by a multiple nozzle system with the nozzles located around the circumference of the riser. The nozzles should be preceded by a catalyst preentry zone operating in the turbulent/fast fluid bed regime.

The nozzle design should provide good atomization as well as good dispersion of the oil droplets across the riser cross section. Since a flat spray pattern will minimize the recontacting of catalyst with fresh oil, this pattern is preferred.

Commercial experience has shown the advantages of this type of system. Table 3-1 shows typical benefits of multiple nozzles over a single nozzle located at the base of the riser[1]. The higher gasoline yield and lower gas yields at higher conversions result from improved catalyst oil contacting and less back mixing of the catalyst in the riser.

Table 3-2 shows the expected effects of a flat spray pattern over an open pipe. The improved performance for this change is

TABLE 3-1. Effect of multiple feed nozzles.

	Change over Single Feed Nozzle
Conversion, Vol. %	Constant
Gasoline, Vol. %	+2.0–4.0
Dry Gas, Wt. %	–0.5–2.0

Schnaith et al., "Advances in FCC Reactor Technology," NPRA Annual Meeting Paper AM-95-36 (1995).

TABLE 3-2. Expected improvement of flat spray vs. open pipe.

Conversion, Vol. %	+1.0 – 2.0
Gasoline, Vol. %	+0.5 – 1.0
Dry Gas, Vol. %	–0.3 – 0.5
Coke, Wt. %	Constant

the result of a more even distribution of the oil across the riser cross section.

The effect of feed atomization has been debated at length. In theory, improved atomization should provide for more rapid vaporization and thus, improved yields.

In practice, this has been hard to verify. Most retrofits to atomizing nozzles have also involved changes in the number and location of the nozzles and/or the nozzle spray pattern.

In addition, it is difficult to assess the actual quality of oil atomization in a commercial riser. Nozzle atomization tests are normally conducted using water and air with the nozzle discharging to the atmosphere. This differs significantly from a commercial FCC operation where the oil and steam mixture is discharged into a riser in the presence of a large mass of catalyst particles.

Recently, however, data has become available from a nozzle revamp where the only major change was in the quality of atomization in the feed nozzles (Table 3-3).[2]

The data are, however, somewhat confused by the fact that the post revamp feed was significantly poorer than that before the

TABLE 3-3. Effect of improved atomization.

	Base	Improved Atomization
Feed Gravity, Degees API	23.2	22.9
Feed Sulfur, Wt. %	0.3	0.37
Feed Nitrogen, WPPM	1247.0	1367.0
Feed Carbon Residue, Wt. %	2.7	3.2
Gasoline, Vol. %	62.5	63.7
LCO, Vol. %	12.1	11.2
Slurry, Vol. %	5.4	5.6
LPG, Vol. %	30.3	30.0
Coke, Wt. %	7.5	8.0
Dry Gas, Wt. %	3.2	2.8

Daria, D. et al. "Advances in Residual Oil FCC," JPI Petroleum Refining Conference (1992).

modification. In addition, the operation of the unit resulted in higher conversion and thus higher coke yields. If the results were to be corrected back to constant operating conditions and constant feed quality, one would expect about a 1.0 volume percent increase in conversion at constant coke (author's estimate). Most of this extra conversion would appear as gasoline and the yield of dry gas would decrease due to reduced thermal cracking.

Commercially Available Feed Injection Systems

Various feed injection system configurations are available from FCC licensors. Probably the best known is the injection system licensed by Stone and Webster Engineering Corporation.

The injection system consists of multiple atomizing nozzles located around the riser circumference. In the nozzle (Figure 3-1), the oil passes through a restriction orifice. The high velocity oil jet from this orifice impinges on a hardened block and this impact serves to atomize the oil. Steam injected at the base of the nozzle provides additional atomization and transports the oil droplets to the nozzle tip. Additional pressure drop at the tip provides additional atomization. The tip is a simple slot designed to produce a flat spray pattern.

98   Chapter 3

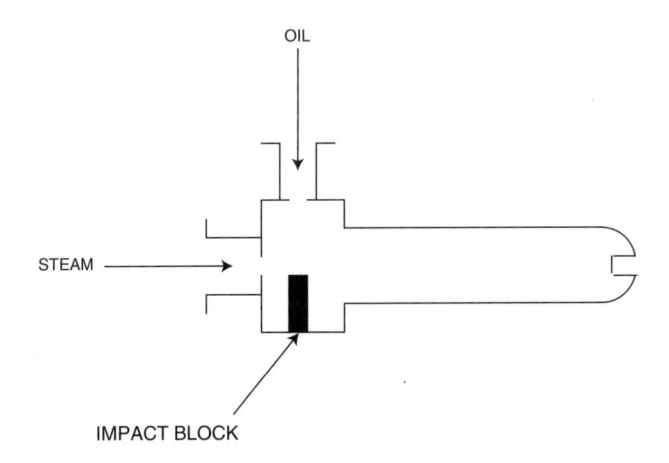

FIGURE 3-1. Stone & Webster nozzle. (Redrawn from Hunt, D.A., "Stone & Webster FCC Feed Injection Technology, Advances and Advantages," *Eighth Annual Refining Seminar,* Stone & Webster Engineering Corporation [1996]. By permission of Stone & Webster.)

The nozzles are located some distance above the base of the riser. Catalyst entering the riser is first fluidized by steam. Thus the catalyst is flowing up the riser as it enters the feed injection zone.

The primary goal of the Stone and Webster injection system is to produce small oil droplets (circa 60 micron diameter). As discussed earlier, small droplets will give rapid vaporization in the riser and produce improved yields.

The introduction of the Stone and Webster nozzles was extremely well-received and many of these systems were retrofitted to older FCCUs.

UOP's first improved nozzle design, the Premix nozzle, was developed for use in the UOP RCC process. These nozzles were used in conjunction with the UOP gas lift feed injection system.

The Premix nozzle used steam atomization to create an oil mist which was then injected into the riser.

In the gas lift injection system, multiple Premix nozzles are placed around the riser circumference. Catalyst enters the riser below the feed injection zone and is carried to the feed nozzles by fuel gas recycled from the gas concentration unit. In addition to transporting the catalyst up to the feed nozzles, UOP reports that this lift gas passivates any nickel on the catalyst and thus reduces the net yield of dry gas from the unit. The lift gas system also minimizes the use of steam at the base of the riser and thus, may reduce hydrothermal deactivation of the catalyst. The gas is, however, a recycle stream that travels through the main fractionator, wet gas compressor, and gas recovery section. Thus, it will increase the load on these sections of the FCC.

In the RCC, the feed nozzles are located well up the riser. Thus in this unit, the catalyst must be accelerated to dilute phase transport velocities to avoid excessive pressure loss in the lift zone. In other lift gas applications, the nozzles are generally much closer to the base of the riser and the lift gas velocities can be lower. UOP has found that higher velocities below the feed nozzles improve yields in one test, increasing velocities by a factor of three, which increased conversion by almost two percent.[1] This effect was attributed to increased penetration of the oil mist into the riser. Catalyst densities in this test were reported to be 15–30 lb/ft^3 which indicate that the preentry zone was still in the dense phase.

UOP also used the Premix chamber design in replacement nozzles for older units. In these units, the existing shower head distributor is replaced with a Premix nozzle in the base of the riser. This nozzle retains the shower head design to disperse the oil mist in the riser.

UOP's current nozzle design is the Optimix nozzle (Figure 3-2). In this nozzle, the oil flows through a central pipe. Steam flows through the annular space between the oil tube and the nozzle body. Proprietary internals contact the steam and oil, and the pressure energy in the steam is used to atomize the oil. The oil/steam mixture flows to the nozzle tip which consists of an array of small orifices oriented to produce a flat spray pattern.[3]

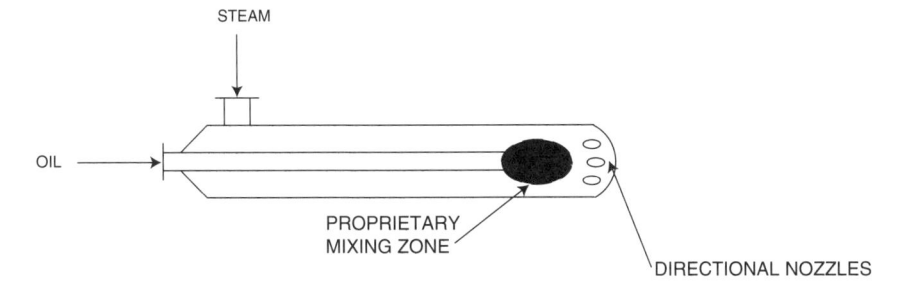

FIGURE 3-2. UOP Optimix nozzle. (Optimix is a trademark of UOP. Dahlstrom et al. "FCC Reactor Revamp Project Execution and Benefits," NPRA Annual Meeting Paper AM-96-28. Used by permission of UOP.)

The Optimix nozzle gives substantially better atomization than the Premix nozzle.

The Lummus Micro-Jet nozzle also uses proprietary internals to mix oil and steam and to atomize the oil. Multiple nozzles are used and these are installed around the riser circumference. The Micro-Jet nozzle uses a multiple orifice array to distribute the oil mist into the riser.[4]

The orifice arrays used in both the UOP Optimix nozzle and the Lummus Micro-Jet nozzle produce a flat spray pattern similar to that produced by a slotted tip. The licensors claim a more even distribution of the oil across the spray for the orifice system.

M.W. Kellogg pioneered the concept of improved spray patterns. Kellogg's earliest efforts in this regard led to the use of slotted tip nozzles. These nozzles, which produced a flat spray pattern, were used to replace the open pipe nozzles used in earlier multiple nozzle systems. The superior performance of the flat spray pattern developed by these nozzles led to its eventual adoption by all major FCC licensors.

The M.W. Kellogg Atomax (Figure 3-3) nozzle is a recent development which replaces their earlier low pressure drop slotted tip nozzles. In the Atomax nozzle, high velocity steam jets are injected into an oil pipe located in the center of the nozzle. These high velocity jets partially atomize the oil and accelerate the oil/steam mixture. Nozzle internals then complete the atomization of the oil. The nozzle tip is a simple slot designed to produce a flat spray.

In the Kellogg system, multiple nozzles are located around the riser circumference. Steam is used to fluidize the catalyst below the feed nozzles.

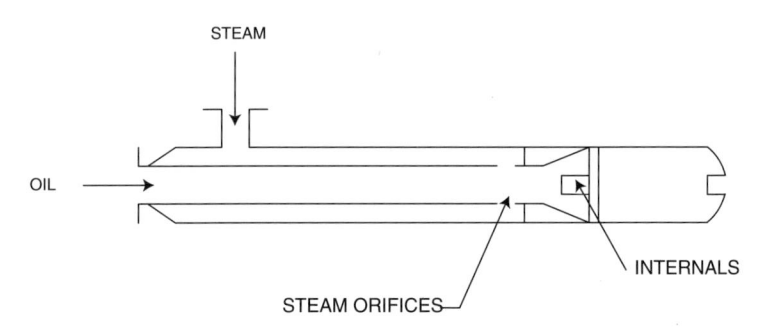

FIGURE 3-3. M.W. Kellogg Atomax nozzle. (Adapted from U.S. Patent No. 5,306,418. Atomax is a trademark of M.W. Kellogg.)

Riser

Following the feed injection section, the catalyst and oil (now vaporized) interact and the cracking reaction takes place. In modern FCCUs, the riser is a vertical pipe. In some older units the riser(s) may be curved.

Typical riser designs call for velocities at the base of the riser to be 10–30 ft/sec and for velocities at the top of the riser to be 40–100 ft/sec. Since the cracking reaction involves the breaking of a large molecule into two or more smaller molecules, there is molar expansion and thus an increase in the volume of gas over the riser length. In some cases it may be necessary to increase the diameter of the riser to maintain the design limits. In a typical FCCU, the oil vapor volume expands by a factor of 3–4 in the riser. Higher conversion operations give higher volume expansions.

Commercial cracking reactions appear to be second order thus most of the reaction takes place in the lower section of the riser. A generally used rule of thumb is the one-third/two-thirds rule. This states that two-thirds of the conversion will take place in the first one-third of the riser volume. Using this rule of thumb, a curve relating the fraction of the total conversion to the fraction of the riser volume can be created (Figure 3-4).

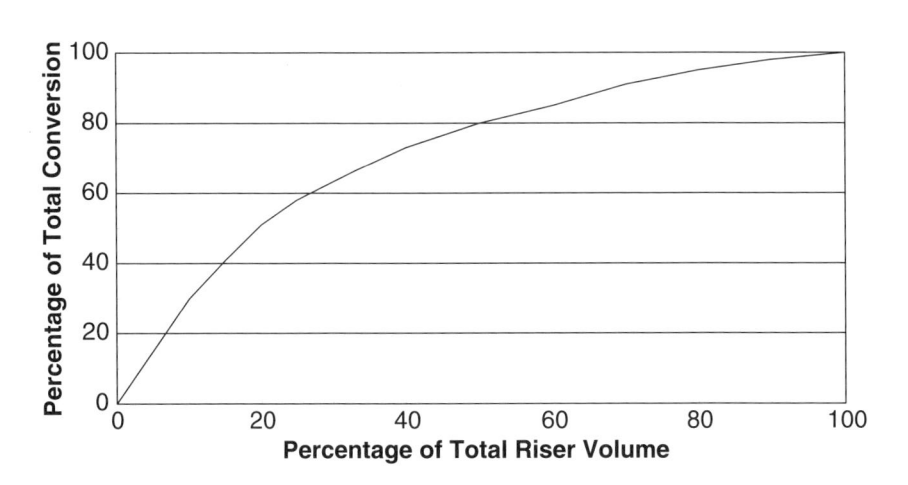

FIGURE 3-4. Molar expansion in riser.

Riser designs are either internal (Figure 3-5) or external (Figure 3-6) to the reactor vessel. The internal design has the advantage of minimizing any horizontal flow in the riser. The external designs allow the use of a cold wall design over the entire riser length.

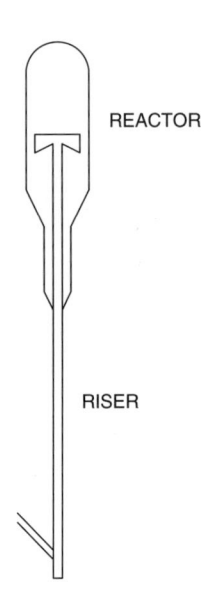

FIGURE 3-5. Internal riser.

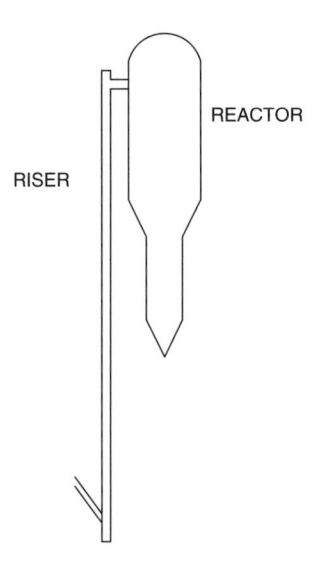

FIGURE 3-6. External riser.

External riser designs require that the catalyst/oil vapor flow turn from vertical to horizontal. The preferred means of making this change in direction is a hard tee (Figure 3-7) rather than an elbow.

The hard tee turn provides a cushion of low velocity catalyst at the top of the tee. This cushion reduces the erosion of the riser at this point. Conventional hard tee turns often do show erosion as indicated in Figure 3-8. This problem can be eliminated by modifying the turn to reduce the velocity. M.W. Kellogg has successfully used a patented version of this modification for several years.

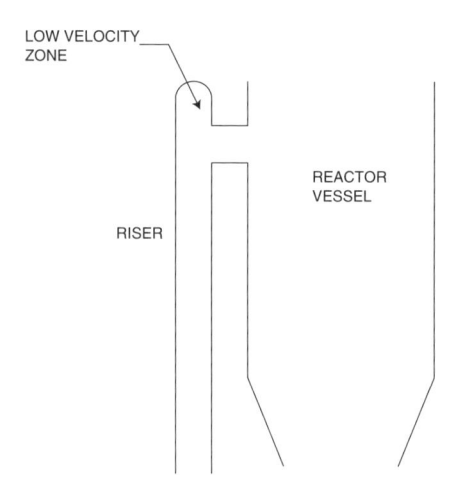

FIGURE 3-7. Hard tee riser turn.

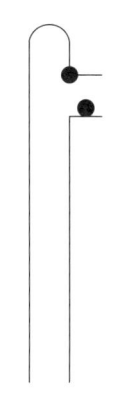

FIGURE 3-8. Hard tee erosion points.

Riser Termination

At the end of the riser, it is important to make a quick separation between the catalyst and the product vapors. This will minimize cracking reactions in the reactor vessel. Due to the low velocities in the reactor, any catalyst present will be in a fully backmixed state. In addition, the catalyst in the reactor, having passed through the riser, will have considerable levels of carbon deposited on the active sites. These two factors combine to give reactor dilute phase cracking poor product selectivities.

In many older units, the risers were intended to serve as a transfer line to bring the catalyst and oil vapors to the reactor bed. In these units the risers had either no terminal separation device, or were equipped with a very simple device designed to distribute the catalyst and oil into the reactor bed. Examples of these simple termination devices were the dragon head separator used in Texaco FCCUs or the duck bill separator (Figure 3-9).

The earliest improvements on these devices were inertial separators. These consisted of devices which forced the catalyst/oil vapor mixture to undergo changes in velocity and/or flow direction. Since the denser catalyst particles have more momentum,

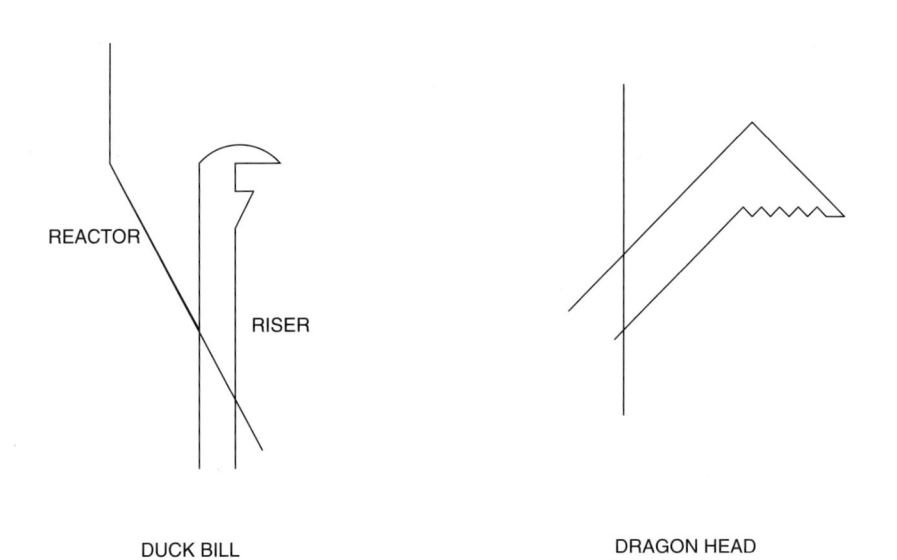

FIGURE 3-9. Simple separators.

they will tend to continue in a straight line and will become separated from the gas steam which will more readily follow a tortuous flow path.

Inertial separators fall into one of two types, simple down turns and cans. The simple down turn (Figure 3-10) discharges the catalyst/oil vapor mixture vertically downward. The discharge is above any reactor bed, but must be below the entry to the riser cyclones. To flow from the riser exit to the cyclone inlets, the vapor and any entrained catalyst must make a 180° change in flow direction. The denser catalyst is less likely to make this turn than the vapor and thus, falls to the reactor bed.

Riser cans have been used extensively on units with internal risers. The simplest can type separator consists of a concentric can placed over the open end of the riser (Figure 3-11). Catalyst and vapors leaving the riser enter the can and flow down the annulus between the can and the riser. The bottom of the can is open and located above the reactor bed. The gas leaving the can turns 180° to flow upward to the cyclones and the catalyst falls to the reactor bed. The simple can is in essence a variation on the down turn separator and shows similar separation efficiencies.

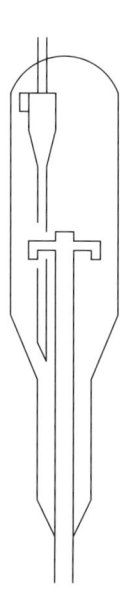

FIGURE 3-10. Down turn RTD.

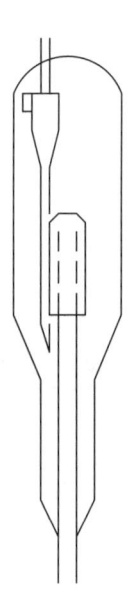

FIGURE 3-11. Can separator.

Down turn and can separators both direct the vapor discharge downward into the reactor vessel. Originally, the vapor was expected to quickly turn upward and exit the reactor through the reactor cyclones. Tracer studies of units equiped with this type of terminator device showed, however, that the vapor flow continued down to the stripper bed. This led to long vapor residence times and thus, increased dilute phase cracking.

Figure 3-12 shows two variations on the simple can separator. These modifications are intended to improve the separation efficiency and/or to reduce the vapor residence time in the reactor.

The first of these is a separator design patented by Chevron. This separator makes use of a frustroconical baffle located below the can outlet. This baffle is followed by a second baffle attached to the reactor or stripper wall. The effect of these two baffles is to force the riser discharge through a rapid 180° turn in the close proximity of the baffle surfaces. In this turn, the solid catalyst tends to contact the baffle surfaces and is separated from the gas stream.

The next variation on the can separator was patented by Stone & Webster. In this design, the can contains a section where the can walls diverge in the direction of flow. Openings in the walls of

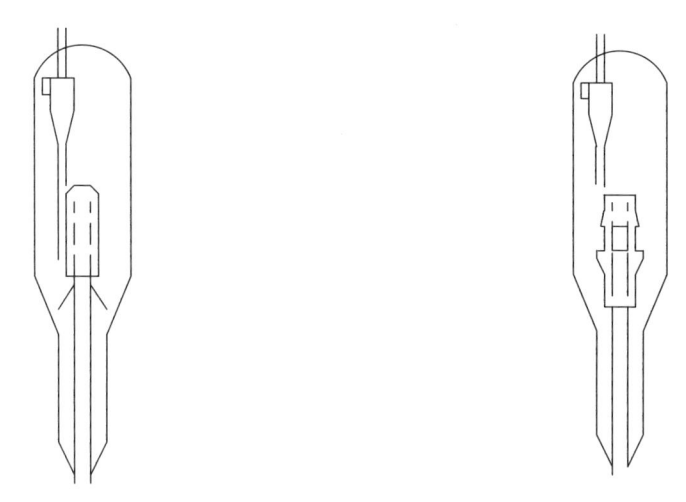

FIGURE 3-12. Modified can designs.

this diverging section serve as exits for the product vapors. The bottom of the can is open and submerged in the reactor bed.

In operation, vapors entering the diverging section of the can slow down and turn to exit through the openings. Catalyst, having a higher momentum, continues in a downward direction and contacts the walls of the can below the diverging zone. This catalyst flows down the walls and exits the can through the open bottom. The diverging section acts to slow the gas flow and improves collection efficiency.

The most unusual variation of the can design is the UOP vented riser (Figure 3-13). In this design the reactor vessel serves as the can. The riser discharges through an open top into the reactor vessel at an elevation above the reactor cyclone inlets. The cyclones themselves are located so that the inlets are in close proximity to the riser. Vapors leaving the riser turn 180° to enter the cyclone inlets. The catalyst leaving the riser continues to flow in an upward direction and contacts the top head of the reactor. The catalyst then flows down the reactor walls into the reactor bed.

Since the vented riser design discharges above the reactor cyclone inlets, there is a constant flow of catalyst and vapors in this section of the reactor. This eliminates a dead zone that exists in other reactor designs and minimizes problems with coking above the reactor cyclones.

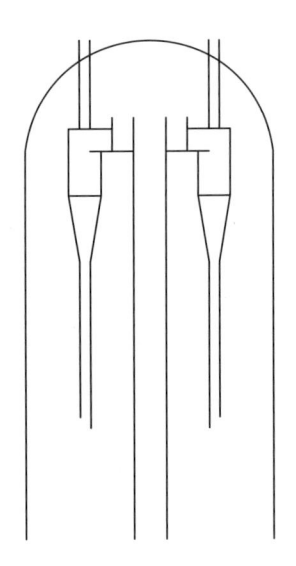

FIGURE 3-13. UOP vented riser.

The fact that the catalyst flows down the reactor walls can lead to erosion of any equipment attached to these walls or projecting into the reactor. Possible areas of erosion include cyclone dipleg support clips, thermowells, and pressure taps. These projections should be protected.

The most recent advancements in riser termination involve the use of cyclones attached to the riser outlet. In the 1970s, UOP and M.W. Kellogg both began to use riser cyclones. At that time, it was thought that the efficiency of the cyclones would be low and the term rough cut cyclones was coined to describe these installations.

Riser cyclones typically have separation efficiencies of 95–98 percent. In most installations, the cyclones operate at a pressure slightly above the reactor pressure. For this reason, the riser cyclone diplegs must be sealed in the reactor bed to prevent gas from flowing down the dipleg into the reactor. Riser cyclone pressure drops can be calculated by the method given in Chapter 6.

The current state of the art in riser terminations is closed cyclones. These systems minimize reactor dilute phase residence time by connecting the riser cyclone outlets directly to the inlets of the reactor cyclones. In essence, the product vapors never enter

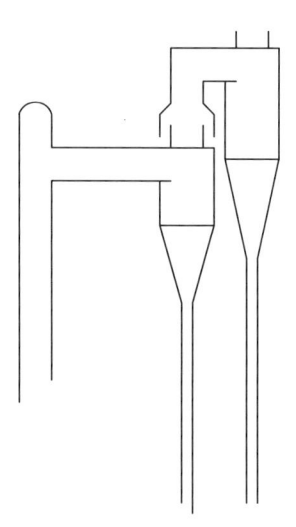

FIGURE 3-14. Mobil closed cyclone system.

the reactor but flow directly from the riser into the cyclones and out of the reactor.

Catalyst from the cyclones flows directly into the reactor bed. This catalyst flow does, however, carry some product vapors into the bed. These vapors are recovered in the stripper. For a closed system to function properly, some means for these stripper vapors to exit the reactor must be supplied.

The closed cyclone system patented by Mobil (Figure 3-14) connects the cyclones by a slip joint located between the riser cyclone outlet duct and the upper cyclone inlets. The annular space in this slip joint is sized to allow the stripper vapors to enter and pass on to the upper cyclones. The first cyclone in the series operates at a higher pressure than the reactor and thus, the diplegs must be sealed to prevent a gas underflow.

The system patented by Texaco (Figure 3-15) uses an annular space between the riser cyclone body and the riser cyclone outlet duct to allow stripper vapors to enter the cyclones. This design differs from the Mobil design in that the riser cyclones operate at a pressure lower than the reactor pressure. This eliminates the need to submerge the diplegs in the reactor bed. Instead, the cyclones operate in much the same manner as a conventional two-stage cyclone system.

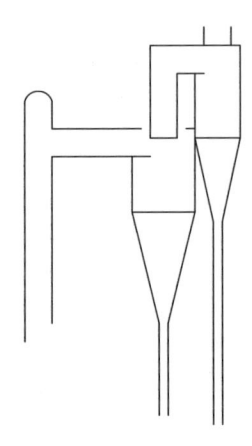

U.S. PATENT 5,248,411

FIGURE 3-15. ABB Lummus closed cyclones.

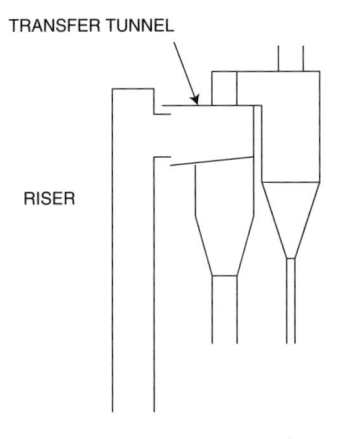

FIGURE 3-16. Exxon closed cyclone system. (Adapted from Ladwig, P.K. et al., "Recent Operating Experience with Flexicracking Commercial Developments in Short Contact Time Catalytic Cracking," NPRA Annual Meeting Paper AM-94-42 [1994].)

In the Exxon design (Figure 3-16), a conventional two-stage cyclone system is located in close proximity to the riser. Product vapors leave the riser through ducts that enter the first stage cyclone inlets. The cyclone inlet is modified to stabilize the vapor flow entering the cyclone body. Stripper gases enter the first stage

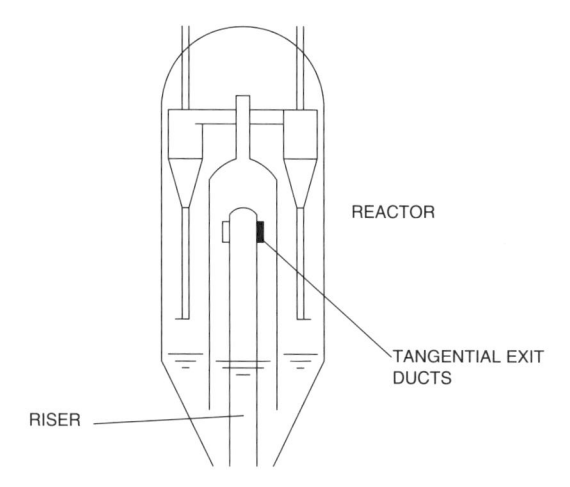

FIGURE 3-17. UOP VSS separator. (Redrawn from Reference 1. Used by permission of UOP. VSS is a Service Mark of UOP.)

cyclone through the space between the riser discharge duct and the cyclone inlet. As with the Texaco system, both cyclones operate at a lower pressure than the reactor.

UOP's VSS separator design (Figure 3-17) uses an internal can located over the riser. Product vapors exit the riser through curved ducts. These ducts are oriented to produce a swirling flow inside the can. This causes the catalyst particles to migrate to the can walls where they are separated from the gas stream. The gas flows out of the top of the can and directly to the reactor cyclones. Small slots in the can wall allow stripper vapors to exit the reactor. UOP has proposed an extension of the VSS design that would eliminate the reactor vessel.[1] In this design, the separator can becomes the pressure vessel. Gases leaving this can flow into external cyclones mounted above the separator.

The advantage of closed cyclone systems is reduced dilute phase cracking and thus, a better yield structure. Data published by Mobil are shown in Table 3-4. These data are based on the actual results after installing closed cyclones on three FCCs.

Since closed cyclone systems minimize the reaction time, they would be expected to reduce the slower secondary reactions such as hydrogen transfer. Thus, the olefin content of the gasoline and LPG products will be higher.

TABLE 3-4. Effect of closed cyclones.

	Effect of Closed Cyclones
Gasoline, Vol. %	+1.3 – +1.7
Lt. Cycle Oil, Vol. %	+0.8 – +1.5
Dry Gas, Wt. %	–0.9 – –1.0

Avidan et al., "FCC Closed Cyclone System Eliminates Post Riser Cracking," NPRA Annual Meeting Paper AM-90-33 (1995).

Since closed cyclones normally reduce catalyst delta coke, the catalyst circulation will increase. This can present a problem in revamps if the unit is already operating at or near its maximum catalyst circulation. Experience has shown that an increase in reactor temperature may also be necessary to maintain conversion in closed cyclone systems. This may partially offset the reduction in gas yield produced by reduced dilute phase cracking.

The primary disadvantage of closed cyclones is that these systems can be less forgiving in the event of unit upsets. Since there is little holding time in the closed systems, pressure swings (e.g., from a gas compressor upset) are more likely to cause catalyst carryover to the fractionator. In addition, the start up and shut-down of units with closed cyclones must be done with more care than is usually required for units with more conventional riser terminations. These matters can be addressed by a properly designed operating procedure. Operating know-how should be a part of the technology package provided by the closed cyclone licensor.

The MSCC process developed by Barco Industries and licensed through UOP eliminates the riser completely.[5] Catalyst and oil flow directly from the feed injection zone to the catalyst separation device. Thus, the feed injection zone and the catalyst separator are the primary reaction zones.

Reactor Vessel

The reactor vessel today serves primarily as a cyclone containment vessel and as a place to collect the spent catalyst that has been separated from the product vapors. Ideally, there will be no reaction in this vessel. Generally, the vessel diameter is set to give a superficial velocity of 1–3 ft/sec based on the combined flow of product vapors leaving the reactor and stripper vapors leaving the stripper. In some cases, this design parameter is overridden by the need to provide adequate room for the reactor cyclones.

Reactor length in older bed cracking units was set to give adequate bed volume plus sufficient height for catalyst disengagement above the bed. These needs generally assured adequate length to accommodate the reactor cyclones. In newer units, designed to operate with efficient riser separators, the reactor length is set by the length of the desired cyclone system.

Reactor vessels should be equipped with an anticoking (dome) steam ring (Figure 3-18). This ring is located in the reactor above the reactor cyclones. The steam flow through the ring should be designed to displace the reactor volume above the cyclones once every three minutes.

In the absence of this ring, the hydrocarbon vapors in this section of the riser are subject to long residence times at reactor temperature. This leads to thermal cracking and eventually to the formation of coke. Over the course of a typical run, the coke buildup in this area can be significant.

These coke deposits can cause several problems. Any thermal cycling of the reactor (such as a short-term feed outage) may

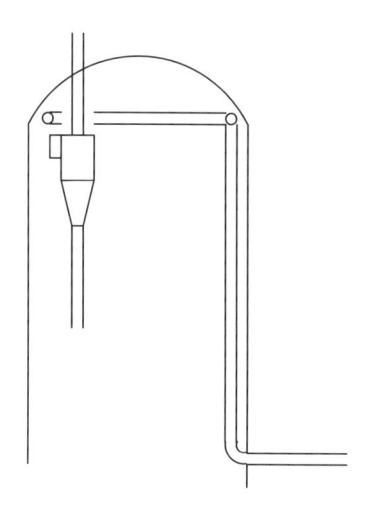

FIGURE 3-18. Anticoking steam ring.

cause some of this reactor coke to break loose. Large chunks of this dislodged coke can fall into the stripper and the spent catalyst standpipe and interfere with catalyst circulation. Large coke deposits can also cause equipment damage when they break loose.

On shut-downs, this coke may become a personnel hazard. As the shut-down reactor cools, portions of the deposit may become partially detached. These loosened portions can later break free and fall. There have also been cases of reactor coke deposits igniting when air is introduced to the shut-down reactor.

Once coke forms in the reactor dome, it is difficult to remove. Generally, the deposits must be removed mechanically using power tools, hammers and chisels. Other methods such as blasting and water cutting have been tried, but often these methods cause extensive damage to the reactor and internals.

The steam used in the dome steam ring must be dry and superheated. Injecting wet steam into the reactor will be counterproductive and will probably lead to increased coking problems. To assure that the steam meets this criteria, the steam line should enter the reactor well below the elevation of the ring as shown in Figure 3-18. This will allow the steam to heat up as it travels through the internal steam line to the ring.

The use of anticoking steam is even more critical in closed cyclone systems. In these systems the small volume of hydrocarbon vapor entering the reactor from the stripper must be displaced from the reactor vessel. In the absence of sufficient anticoking steam these vapors will remain in the reactor for a very long time. Some early closed cyclone systems lacked proper anticoking steam systems and these units suffered massive coke buildups in the reactor.

Both hot wall and cold wall reactors have been used in FCCUs. Hot wall reactors are made of alloy and are externally insulated. During operation, the reactor wall is at the reactor temperature. Hot wall reactors may have a thin layer of erosion resistant refractory on the inside of the reactor.

Cold wall reactors are made of carbon steel with internal insulation. This insulation is generally in the form of a medium weight refractory which is gunned onto the reactor wall. There may also be an erosion resistant lining on the inside surface of the reactor lining. This is usually limited to potential areas of high velocity vapor impingement.

Cold wall reactors are less subject to metallurgical problems and are preferred for new installations. There may also be cases

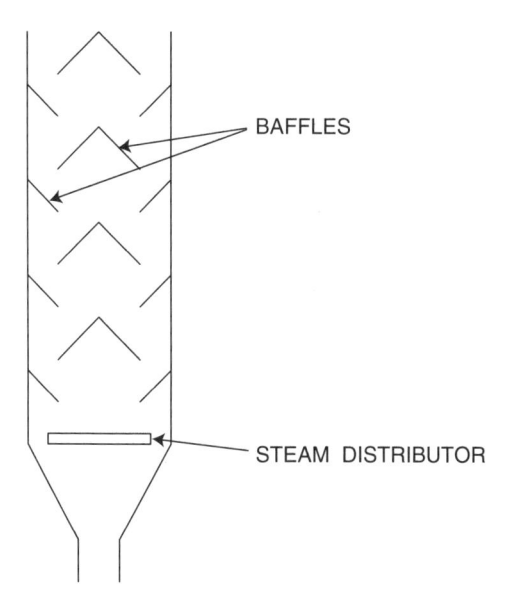

FIGURE 3-19. Flow in unbaffled stripper.

where converting an existing hot wall reactor to cold wall will be desirable.

Catalyst Stripper

The catalyst leaving the riser/reactor system will contain significant quantities of entrained hydrocarbon vapors. If these vapors are not removed from the catalyst, they will flow into the regenerator and be burned as coke. This can result in high catalyst regeneration temperatures. Hydrocarbon vapors entrained into the regenerator also represent lost products.

The catalyst stripper removes most of these entrained hydrocarbons by displacing them with steam. In a typical stripper (Figure 3-19), the catalyst flows downward over a series of baffles. Steam is injected through a ring at the bottom of the stripper and flows up through the catalyst bed.

Typical design parameters for the catalyst stripper are:

Catalyst superficial mass velocity	600–1000 lb/ft^2/sec
Free area between baffles	45–55% of stripper area

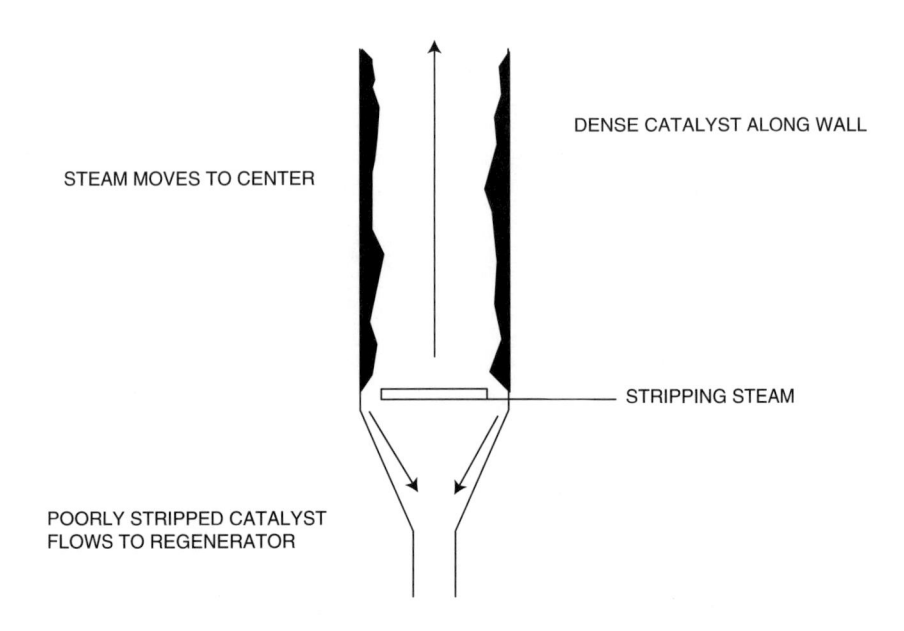

FIGURE 3-20. Catalyst stripper.

Catalyst residence time   0.5–1.0 minutes
Superficial steam velocity   0.5–1.0 ft/sec

Baffleless strippers should be avoided. In a stripper without baffles, dense catalyst will tend to collect near the vessel walls and the steam will tend to move up the center of the vessel (Figure 3-20). This will result in poor catalyst and steam contacting and reduce stripping efficiency.

Baffles can be either disk and donut or shed decks (Figure 3-21). Shed decks generally require higher maintenance, but were long thought to give better steam distribution than disk and donut designs. In fact, a properly designed disk and donut baffle system can provide steam distribution as good as a shed deck system.

Problems with steam distribution in disk and donut baffles occur when the bottoms of the baffles are not level. This may be due to poor installation, or it may be the result of some warpage of the baffle during operation. If the baffle is not level, the steam will not flow evenly around the baffle, but will instead flow preferentially over the higher side (Figure 3-22). To eliminate this

DISK AND DONUT

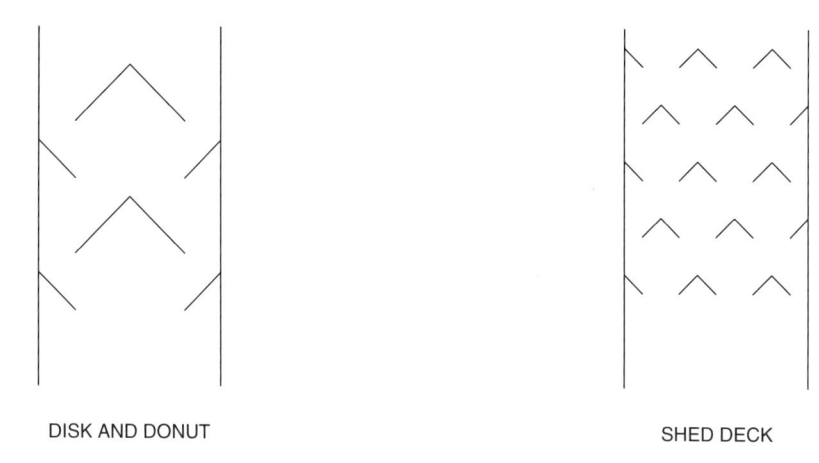

SHED DECK

FIGURE 3-21. Baffle types.

IF BAFFLE IS NOT LEVEL, STEAM FLOWS FROM UNDER HIGH SIDE

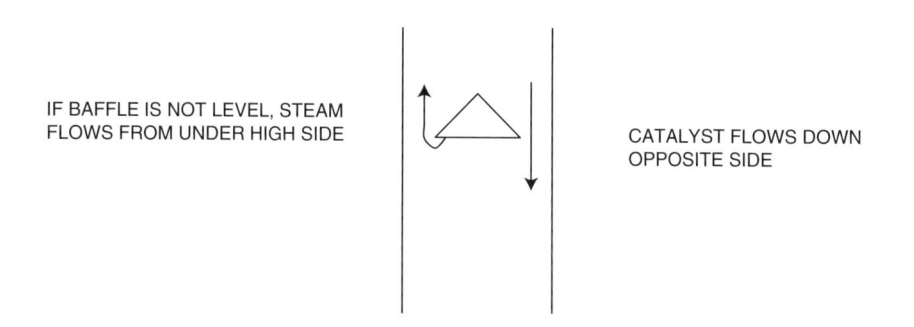

CATALYST FLOWS DOWN OPPOSITE SIDE

FIGURE 3-22. Flat bottom baffles can cause poor steam distribution.

possibility, disk and donut baffles should be designed with a notched bottom and/or with a series of holes in the baffle to distribute steam (Figure 3-23).

The sizing and placement of the holes in the baffles is critical to proper operation of the stripper. Too many holes placed too high up the baffle will result in all or most of the steam flowing along the side of the stripper and/or through the center of the stripper. This will reduce the contact of steam and catalyst and impair stripper performance.

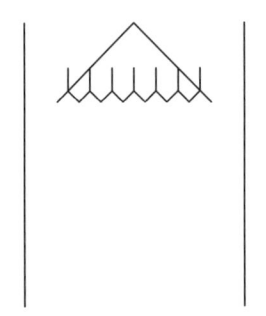

FIGURE 3-23. Notched bottoms or holes in baffles improve steam distribution.

A typical design would have 25% of the stripping steam flowing through each set of baffle holes. The holes are generally 1/2–1 inch in diameter and are located 25% up the baffle slope. To avoid coalescence of the bubbles from the holes, the minimum spacing between the holes should be six inches.

The actual number of holes can be set based on the desired steam flow and the available pressure drop through the holes. The available pressure drop is equal to the vertical height of the holes above the bottom of the baffle times the stripper bed density. The velocity through the holes can then be calculated by:

$$V_{hole} = C_D \sqrt{\frac{(144)(2)g_c \Delta P}{\rho_{stm}}}$$

where:

V_{hole} = velocity of steam through hole, ft/sec
C_D = coefficient of discharge = 0.82
g_c = 32.2
ΔP = pressure drop through hole, psi
ρ_{stm} = steam density at stripper conditions, lb/ft^3

The volume flow of steam is then equal to:

$$ACFS = V_{hole} \times \text{total hole area}$$

and the mass flow of steam is:

$$Lb/hr = ACFS \times \rho_{stm}$$

STEAM RING   PIPE GRID   STEAM CHEST

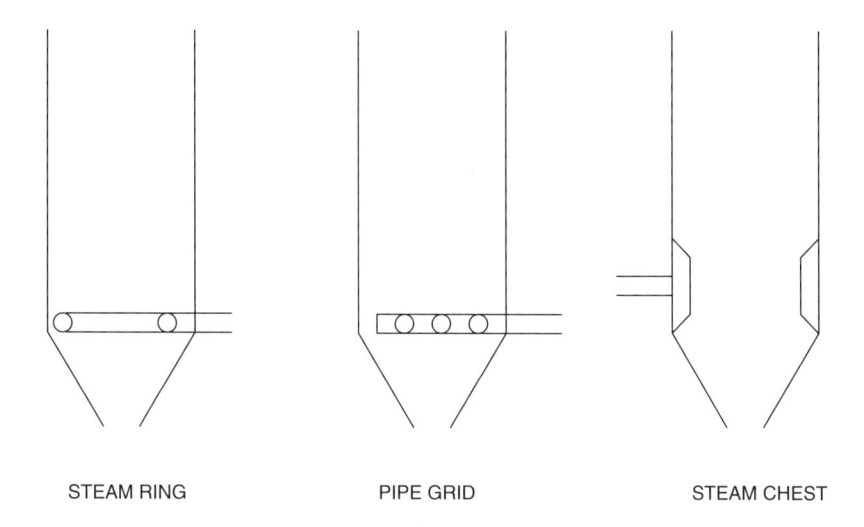

FIGURE 3-24. Stripping steam distributors.

Stripping steam distributors can be in the form of rings, spiders (pipe grids) or steam chests (Figure 3-24). Rings or spiders are preferred as steam chests are more likely to experience mechanical failure during operation.

A well-designed distributor system will assure that the steam covers the entire stripper cross section so that steam will collect under all of the baffles. Pipe grids do this automatically. For steam ring systems, the nozzle orientation and ring placement must be such that steam flows to each set of baffles.

Many current stripper designs include prestripping. This is done by locating a steam distributor at the top of the stripper. Steam injected at this location serves to remove the most volatile hydrocarbons at the top of the stripper. This prevents these products from cracking to gas and coke in the stripper bed. Prestripping steam is generally set by experimentation to give the best stripping efficiency.

Strippers with larger length to diameter ratios (L/Ds) tend to be more efficient. This accounts for the observed improved efficiency of annular strippers (Figure 3-25). All else being equal, an annular stripper has a lower effective diameter and thus, a higher L/D.

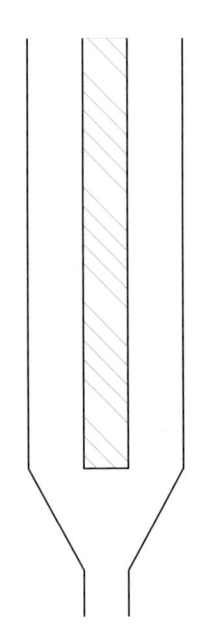

BAFFLES NOT SHOWN FOR CLARITY

FIGURE 3-25. Annular stripper.

Operating Considerations

Operational monitoring of the riser/reactor system requires an understanding of the basic design of the feed nozzles, riser, and stripper. Each of these systems will have been designed to operate within a specific range. In many cases, operation outside of these limits will appear to offer benefits in terms of reduced costs or greater operating flexibility. Often, however, these apparent advantages are gained at the price of accelerated equipment damage or longer term operational difficulties. This section contains general guidelines for the operation of the reaction system.

Feed Nozzles

Proper operation of atomizing feed nozzles requires attention to both the oil and steam flow rates. Nozzle atomization is dependent on the proper mix of oil to steam and on maintaining the proper pressure drops and velocities in the nozzle. The nozzle supplier's recommendations regarding minimum flows and turn-down procedures should be followed exactly.

Many of the modern nozzle designs require fairly high dispersion steam rates. This results in increased production of sour water from the FCC. FCC engineers are often tempted to reduce nozzle steam in an attempt to reduce operating cost and/or sour water production. This practice should be avoided.

If the nozzle steam is reduced below the minimum required for good atomization, the yields on the FCC will suffer. The lost revenue from poorer yields will more than offset any savings in operating costs. In addition, poor atomization in the riser can lead to excessive coking in the riser and the transfer line. This can lead to an early shut-down or to an extended turn around.

Risers

The primary concerns in riser operation are: 1) the choke velocity of the riser; 2) the riser pressure drop; and 3) the catalyst hold-up in the riser. These three issues are related. As the catalyst mass in the riser increases, the catalyst hold-up, the riser pressure drop and the choke velocity all increase. There are many different correlations for each of these riser parameters. Those presented here were originally published by Knowlton et al.[6] experimenting with powdered coals.

Choke Velocity. For any given riser catalyst mass velocity, there is a minimum vapor superficial velocity below which the riser will not remain in dilute phase transport. Below this velocity, the catalyst/vapor mixture drops abruptly into a dense phase. The result is a sudden increase in riser pressure drop. When this occurs, the riser is said to choke and the velocity at which this transition occurs is known as the choke velocity.

In FCC operations, the design of the riser generally assures that the normal operation is well above the choke velocity. During start up or shut down, however, velocities in the riser are generally low, and choking can occur if the catalyst mass velocity is too high.

Start up operations are particularly susceptible to choking. During start up, steam is generally used to lift hot catalyst from the regenerator to the reactor. The flow of steam is set at a predetermined level and the flow of catalyst is regulated manually. The quantity of steam that can be injected into the riser is often limited by injection system velocity considerations, and the riser velocities during steam circulation are generally much lower than the normal operating velocities.

Risers can choke during start up circulation if too much catalyst is introduced to the riser. Once the riser chokes, it is generally necessary to withdraw catalyst from the riser before it can be returned to dilute phase operation.

The problem can be avoided by determining the choke velocity curve for the riser. The choke velocity for a given riser and catalyst mass velocity can be estimated by:[6]

$$\frac{V_{ch}}{\sqrt{gd_p}} = 9.07 \left(\frac{\rho_p}{\rho_g}\right)^{0.347} \left(\frac{Wd_p}{\mu}\right)^{0.214} \left(\frac{d_p}{D_T}\right)^{0.246}$$

where:
 V_{ch} = choke velocity, ft/sec
 g = gravitational constant, 32.2 ft/sec^2
 d_p = average particle size, ft.
 ρ_p = particle density, lb/ft^3
 ρ_g = gas density, lb/ft^3
 W = solids mass velocity, lb/ft^2-sec
 μ = gas viscosity, lb/ft-sec
 D_T = riser diameter, ft.

Using this equation and the known properties of the catalyst and gas, a curve of choke velocity versus catalyst mass velocity can be constructed. As long as the actual velocity in the riser is above choke velocity, the riser will remain in dilute phase.

Riser Pressure Drop. The riser pressure drop consists of three main components. These are the pressure lost in accelerating the feed and catalyst to riser velocities, frictional loss in the riser and static head loss due to the density in the riser. In addition, if the riser includes any turns, there will be additional pressure drop associated with these changes in flow direction.

The overall pressure drop in a riser exclusive of any turns or exit losses is given by the following equation:[6]

$$\Delta P = \frac{\Delta V_g^2 \rho_g}{2g_c} + \frac{W\Delta V_p}{g_c} + \frac{2f\rho_g V_g^2 L}{g_c D_T}\left[1 + \left(\frac{f_p V_p}{fV_g}\right)\left(\frac{W}{V_g \rho_g}\right)\right] + \frac{WH}{V_p} + \rho_g H$$

where:

P = pressure, lb/ft^2
W = solids mass velocity, lb/ft^2-sec
g_c = gravitational constant, 32.2 lb$_m$-ft/lb$_f$-sec^2
V_g = gas velocity, ft/sec
V_p = particle velocity, ft/sec
f = fanning friction factor, dimensionless
f_p = particle friction factor, dimensionless
L = riser length, ft.
D_T = riser diameter, ft.
ρ_g = gas density, lb/ft^3
H = vertical height of riser, ft

The first two terms in this equation reflect the acceleration losses for the gas and the solids respectively. The third term is for the frictional losses in the riser and the last two terms are for the static head from the solids and the gas in the riser.

In order to calculate the pressure drop, two other terms must be estimated. These are the particle velocity (V_p) and the particle friction factor (f_p). Based on experimental data covering a wide range of operations, Knowlton developed the following formulas:[6]

$$V_p = \sqrt[3]{(V_g - V_{ch})V_g^2}$$

and

$$f_p = \left[0.02515\left(\frac{W}{\rho_g V_g}\right)^{0.0415}\left(\frac{V_p}{V_g}\right)^{-0.859}\right] - 0.03$$

Using these correlations, Knowlton was able to fit the pressure drop equation to operations with the following range of variables:

Gas velocity   5–80 ft/sec
Mass velocity   0–150 lb/ft^2-sec

Gas density 0.14–3.4 lb/ft^3
Particle density 78.6–244 lb/ft^3
Particle diameter 0.000525–0.00119 ft.

In applying these correlations to commercial FCC risers, the fact that the volumetric flow rate of the vapor phase and hence, the velocity, increases over the riser length must be addressed. This is normally done by dividing the riser into several short sections and using a curve as shown in Figure 3-4 to estimate the velocity in each section.

Additional pressure losses occur in riser turns and in the exit from the riser. Each of these can be estimated by the following formula:

$$\Delta P = K \left(\frac{V_g^2 \rho_g}{2 g_c} + \frac{W V_p}{g_c} \right)$$

Where the factor K comes from the following table:[7]

	K
45° Turn	0.2
90° Turn (tee)	1.3
Exit (other than riser cyclone)	1.0

For risers which terminate in a riser (rough cut) cyclone, there is no exit loss. The pressure drop through the riser cyclone can be calculated as discussed in Chapter 6.

Riser Terminators

All riser terminators are dependent on velocity to affect solid/vapor separation. Thus, reduced throughput to the FCC will generally result in a decrease in separator efficiency. This can lead to excessive catalyst carryover to the main fractionator.

These problems can be minimized by reducing the reactor pressure and thus, increasing the gas volumes. This can normally be done without difficulty since the total volume of gas at reduced throughput and lower pressures is still generally less than at the normal feed rates.

Many riser termination devices operate at a positive pressure relative to the reactor vessel. The catalyst discharge from these RTDs must be sealed in the reactor bed to prevent gas underflow. The seal depth must be sufficient for the bed static head to overcome the positive pressure in the RTD.

Units equipped with this type of riser terminator must avoid low reactor bed levels as these will lead to gas underflow from the RTD. The increased gas residence time in the reactor from this underflow will result in higher yields of gas and higher catalyst delta coke.

Stripper

Optimum stripper operation is dependent on both stripper design and on the operation of the FCC. Since the stripper is a major consumer of steam, stripper optimization to minimize the steam flow can yield savings in both utility costs as well as reduced sour water rates.

Stripping steam rates are normally 2–4 lb. steam/1000 lb. of catalyst. The optimum rate for a specific unit is a function of both feed and catalyst properties. Heavy feeds and low pore diameter catalysts may require higher stripping steam rates. The optimum steam rate for a particular unit and operation can be determined by watching regenerator temperature and flue gas oxygen content as the stripping steam rate is adjusted.

Reducing stripping steam to increase regenerator temperature should be avoided as this will cause high catalyst particle temperatures in the regenerator and thus lower catalyst activity. In addition, the unstripped hydrocarbons represent lost product. If higher regenerator temperatures are needed, it would be better to introduce slurry recycle or to add a small amount of resid to the feed.

Another measure of stripping efficiency is the hydrogen content of the coke (see Chapter 2). A unit with a well-designed and properly operated stripper will generally produce 7% or less hydrogen in the coke. Values above 8% indicate poor stripping. Hydrogen values below 5.0% are suspect and should be checked carefully.

The hydrogen in the coke is a function of the hydrogen content of the coke deposited on the catalyst (often referred to as hard coke) and the quantity of volatile hydrocarbons entrained from the stripper (soft coke). The hard coke on the catalyst consists of highly condensed aromatic molecules. The lower limit for the hydrogen content of this hard coke is 4.0 wt.%. For most units the hard coke contains between 4.5–6.0 wt.% hydrogen.

The entrained hydrocarbons are essentially reactor vapors. The hydrogen content of these vapors is a function of the feed hydrogen content. Typically, this soft coke contains between 10.0 and 13.0% hydrogen.

The one exception to this general rule is units that operate at very low conversions. Research done in India indicates that the hard coke produced during low conversion cracking contains higher levels of hydrogen.[8] Thus, low conversion units may show high values for hydrogen in coke even though the stripper is operating well.

Stripper performance can also be evaluated through sampling and analysis of the catalyst leaving the stripper. These samples must be taken into a closed container (bomb) to prevent the escape of volatile hydrocarbons. The sample should be cooled to ambient temperatures before being removed from the bomb.

Once the sample is cooled, a portion is analyzed to determine the weight percent of carbon and hydrogen. This value represents the coke on spent catalyst. A second sample, taken in an open container is also cooled and analyzed for carbon and hydrogen. The volatile hydrocarbons are assumed to have escaped from this sample. Thus, the difference in coke between the two samples indicates the volatile hydrocarbons in the sample.

Another alternative is to sample the gas at the bottom of the stripper. This gas sample is then analyzed for hydrocarbon content.

Typically, the volatile hydrocarbons leaving a good stripper will be 10% or less of the total coke on the spent catalyst. Values of 6–7% would indicate excellent stripping. A value of 15% would indicate a stripper with serious operational or design problems.

As the catalyst circulation increases, the catalyst mass velocity through the stripper will increase. This will tend to pull more of the stripping steam down the stripper and into the spent catalyst standpipe. This will have a negative impact on stripper performance.

This is a common problem on older FCCs that are operating well above their original design capacity. In these units a stripper revamp can often improve overall unit performance.

The problem cannot, however, be resolved by simply trimming the stripper baffles to provide more open area. While this will decrease the entrainment of stripping steam, it will also reduce the steam catalyst contacting in the stripper. This will decrease the stripping efficiency and generally offset any improvements due to reduced entrainment.

The problem of excessive catalyst mass velocities can best be corrected by increasing the flow area in the stripper. This has generally required a new, larger stripper vessel. Recently, however, M.W. Kellogg has begun to license a modified baffle design developed by Mobil Oil. This patented design uses downcomers in the stripper baffle to provide additional flow area in the stripper without compromising contact between the catalyst and the stripping steam.

References

[1] Schnaith, M.W., Gilbert, A.T., Lomas, D.A., and Myers, D.N., "Advances in FCC Reactor Technology," NPRA Annual Meeting Paper AM-95-36 (1995).

[2] Johnson, A.R., Dharia, D., and Long, S., "Advances in Residual Oil FCC," JPI Petroleum Refining Conference (1992).

[3] Dahlstrom, B., Ham, K., Becker, M., Hum, T., Lacijan, L., and Lorsbach, T., "FCC Reactor Revamp Project Execution and Benefits," NPRA Annual Meeting Paper AM-96-28 (1996).

[4] Glendinning, R.J., McQuiston, H.L., and Chan, T.Y., "Implement New Advances in FCC Process Technology," *Fuel Reformulation,* March/April (1995), pp. 45–53.

[5] U.S. Patent No. 4,985,136.

[6] Knowlton, T.M. and Bachovchin, D.M., "The Determination of Gas-Solids Pressure Drop and Choking Velocity as a Function of Gas Density in a Vertical Pneumatic Conveying Line."

[7] Perry, R.H., Chilton, C.H., and Kirkpatrick, S.D. eds., *Chemical Engineers Handbook, Fourth Edition,* McGraw Hill (1963), pp. 5–33.

[8] Das, A.K., Singh, H., and Ghosh, S., "Evaluation of FCC Stripper Perfomance: A Fundamental Approach."

4
Regenerator Design

Catalyst leaving the reaction section contains a carbonaceous deposit on its surface. In addition, the catalyst will contain adsorbed and entrained hydrocarbons which were not removed in the stripper. These materials are collectively referred to as coke.

The carbon deposited on the catalyst surface occupies active catalyst sites and thus reduces the available catalyst activity. This deposit, along with other forms of coke, is burned in the regenerator. This restores the catalyst activity and provides the heat necessary for the cracking reactions.

Regenerators are large vessels where the catalyst and air are mixed in a fluid bed. As with any combustion equipment, fuel and air distribution is critical to efficient operation. This basic need is complicated by the fact that the fuel is a solid deposit on the surface of a porous particle.

Coke Combustion

The basic combustion reactions are:

$$2C + O_2 \rightarrow 2CO$$
$$C + O_2 \rightarrow CO_2$$
$$4H + O_2 \rightarrow 2H_2O$$
$$S + O_2 \rightarrow SO_2$$

and

$$2SO_2 + O_2 \leftrightarrow 2SO_3$$

Besides these combustion reactions, there is another reaction that occurs in the regenerator that does not involve free oxygen. This reaction is:

$$C + CO_2 \leftrightarrow 2CO$$

This final reaction can become especially significant in units with poor catalyst mixing. In these units, the formation of carbon monoxide from this reaction can be a major cause of localized afterburning in the regenerator cyclones.

The regenerator system consists of the regenerator vessel, air distributor, catalyst distributor, if any, cyclones, and plenum. Additional equipment outside of the regenerator proper are the air blower and the air heater.

Regenerators can operate in partial or complete combustion modes. Early units operated in partial combustion. In these units, the flue gas leaving the regenerator contains little or no excess oxygen and a large volume of carbon monoxide. Normally about half of the carbon burned in a partial burn unit leaves the regenerator as CO.

In complete combustion operation, sufficient excess air is supplied to the regenerator to insure that the carbon is burned completely to carbon dioxide. This technique was originally patented by Amoco.[1] In early complete combustion units, much of the carbon monoxide formed in the bed was burned in the dilute phase. These early complete burn units often showed dense/dilute phase temperature differences of as much as 100°F.

Mobil's development of a platinum based combustion promoter made it possible to shift the oxidation of carbon monoxide from the dilute phase to the regenerator bed. This greatly simplified the operation of complete burn units and eliminated the high dense/dilute temperature differentials.

Today, most FCCs operate in complete combustion. Notable exceptions are UOP RCC units and Stone and Webster/IFP units, both of which have two-stage regenerators. Other FCCs may operate in partial combustion due to regenerator temperature or air blower capacity limits.

Vessel Design

The regenerator vessel is a cold wall internally lined carbon steel vessel. Typically today, the lining is a gunned refractory designed to provide insulation for the vessel shell. Some older units were originally constructed with firebrick lined regenerators. Many of these firebrick linings are still in service.

Vessel dimensions are a compromise between the need to retain catalyst and the need to minimize the capital cost of the regenerator. A large vessel with low velocities could rely on gravity settling to retain catalyst. Such a vessel would, however, have a high capital cost and would require a high catalyst inventory.

By using cyclone separators to collect and return catalyst entrained from the regenerator bed, higher bed velocities can be used. This results in smaller vessels and thus, lower costs.[2] Most regenerators are designed to operate in the bubbling/turbulent bed fluidization regime (see Chapter 7, Fluidization). Typical designs call for bed superficial velocities between 2.0 and 3.0 ft/sec. Turbulent bed regenerators have been operated successfully at up to 4.0 ft/sec.

These regenerators all have a discrete dense phase catalyst bed topped by a dilute phase. Cyclones collect catalyst particles entrained from the bed and return them to the dense phase. Catalyst is withdrawn from the bed and returned to the riser.

Two-stage regenerators have been used since the early 1970s. The earliest designs (Figure 4-1) used an internal baffle arrangement to segregate the regenerator vessel into two separate stages.[3] Catalyst in the first stage was partially regenerated in an oxygen deficient atmosphere. The flue gas from this stage was rich in carbon monoxide.

The partially regenerated catalyst from the first stage overflowed into the second stage where it was contacted with additional air to complete the combustion of the coke deposits. This stage operated with excess air to improve the rate of carbon burning.

The flue gas steams from the two stages were combined in the dilute phase and this proved to be a major problem for these regenerators. When the oxygen-rich flue gas from the second stage mixed with the CO-rich flue gas from the first stage, the carbon monoxide would often ignite. This produced severe afterburning, usually in the cyclones above the second stage.

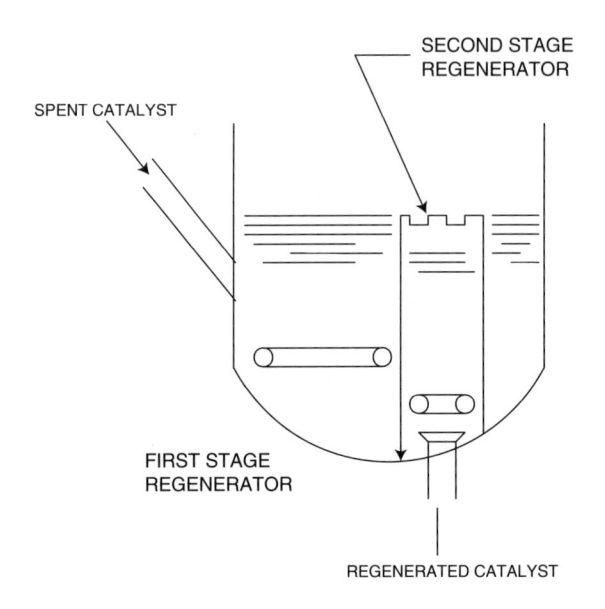

FIGURE 4-1. Early two-stage regeneration.

To correct this problem, it was necessary to reduce the air rate to the first stage until enough carbon entered the second stage to consume all of the oxygen. This largely negated the advantages of the two stage arrangement.

Both the Total resid cracking process (jointly licensed by Stone and Webster and Institut Francais de Petrole) and UOP's Reduced Crude Conversion (RCC) process use two-stage regeneration to reduce the heat released by the combustion of coke. In the RCC process, the first stage is located on top of the second stage. As with all two-stage designs, catalyst is partially regenerated in the first stage in an oxygen lean atmosphere. Catalyst from this stage flows through a standpipe and slide valve to the second stage.

Combustion in the second stage is in the presence of excess air. Flue gas from the second stage flows through a series of vent pipes into the bottom of the first stage catalyst bed. Thus, the excess oxygen in the second stage flue gas is used during the first stage catalyst regeneration. The combined flue gases leave the first stage through a single flue gas system.

In the Total process, the first stage is located below the second stage. The partially regenerated catalyst is lifted from the first stage bed to the second stage via a catalyst lift line. A portion of the second stage air is used to lift the catalyst. The flue gas leaves each stage through a separate flue gas line. Since the flue gas from the first stage contains considerable CO, it cannot be combined with the oxygen containing second stage flue gas until it has been cooled sufficiently to prevent ignition of the carbon monoxide.

The UOP high efficiency regenerator operates in the fast fluid bed regime in the combustor section. The catalyst enters the regenerator at the bottom of the combustor vessel and flows upward along with the regeneration air. Catalyst and air exit at the top of the combustor section and flow up the combustion riser. The upper section of the regenerator serves as a containment vessel for the regenerator cyclones and as a collection vessel for the regenerated catalyst. Catalyst is withdrawn from this upper section and returned to the riser. Hot catalyst from this section is also recycled to the bottom of the combustor section to preheat the spent catalyst from the stripper.

The high velocities in the combustor promote catalyst/air mixing and greatly reduce the catalyst residence time required for regeneration. The result is a smaller regenerator vessel and a greatly reduced catalyst inventory. The excellent mixing in the combustor vessel also accelerates the combustion of carbon monoxide to carbon dioxide. Many high efficiency regenerators do not require the addition of combustion promoter to sustain complete combustion in the catalyst bed.

Regardless of the regenerator configuration, the design regenerator bed velocity usually sets the regenerator diameter. Some regenerators use an enlarged vessel diameter above the bed to improve catalyst disengaging or to accommodate the regenerator cyclones. The regenerator vessel length is set by minimum catalyst disengaging height above the bed and/or the minimum cyclone dipleg length above the bed.

Some earlier regenerators were equipped with antiexplosion or dome steam rings. These rings were designed to displace flue gases from the space above the regenerator cyclones. This was intended to prevent the formation of an explosive mixture of oxygen and carbon monoxide in this dead space. This practice has been discontinued and many of these rings have been removed or taken out of service.

Air Distributor

There are three types of regenerator air distributors in general use: grids, pipe grids (spiders), and rings. Each of these types has both advantages and disadvantages; but if proper attention is paid to design and maintenance, all can give satisfactory performance.

Grids

Grid distributors consist of perforated metal plates which cover all or most of the regenerator cross section. Figure 4-2 shows a flat plate distributor of the type installed in early FCCU regenerators. Both the air and catalyst flowed through these grids. Since the grid covered the cross section of the regenerator vessel, the air and catalyst were well distributed—at least as long as the grid remained in good repair. The fact that the catalyst passed through the grid, however, led to high levels of erosion on the grid holes. This coupled with problems related to the expansion of the grid plate at regenerator operating temperatures resulted in frequent failures of this type of design. These failures led to poor air distribution.

Domed grids in Figure 4-3 were introduced to reduce the problems associated with grid expansion. Many of these early

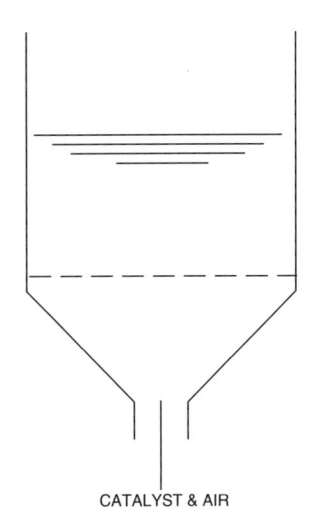

FIGURE 4-2. Flat grid distributor.

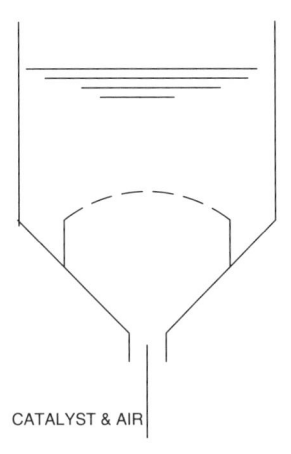

FIGURE 4-3. Domed grid.

grids also had both catalyst and air flowing through the grid holes. Grids of this type gave better service than the flat plate grids. Erosion was still a problem, however. In addition, air distribution suffered. In a domed grid, the pressure drop through the holes in the center of the dome is greater than for the holes on the outside of the dome. Thus, more air flows through these center holes. Most grid designs did not adjust the hole size or density to compensate for this fact, and this led to excessive air in the center of the regenerator and inadequate air near the vessel walls. The natural tendency in fluid beds is for the gas to move to the center of the bed. Domed grids exacerbated this tendency.

Dished grids were used in Esso Model IV units. Catalyst did not flow through these grids. Instead, the catalyst entered via a spent catalyst riser (see Figure 4-4). These designs tended to place more air close to the vessel walls. This compensated somewhat for the tendency of the gas to move to the center of the bed. Dished grids, however, suffered from many of the same expansion-related problems associated with flat grids.

Another problem associated with grids is failure of the grid seals. Flat grids and dished grids are generally supported from the regenerator wall. Since the regenerator is a cold wall vessel, there is considerable difference in expansion between the vessel wall and the hot grid. To allow for this expansion, flexible metal seals were used at the connection of the grid to the vessel. These seals

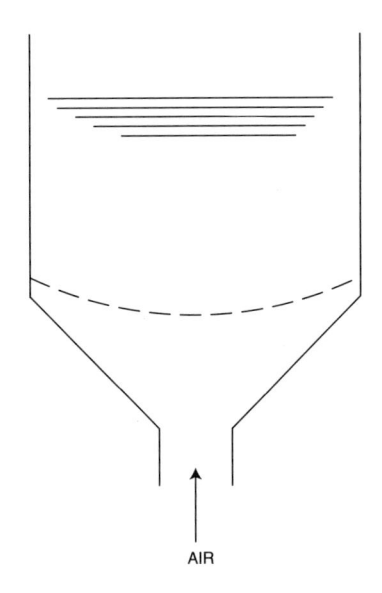

FIGURE 4-4. Dished grid.

were frequently points of failure. Once a seal failed, most of the air flowed through the failed seal and regenerator air distribution suffered. Seals were also required where catalyst standpipes penetrated the grid. These seals were also prone to failure.

Pipe Grids

Pipe grids, or spiders, consist of an arrangement of branched pipes and nozzles designed to cover the regenerator cross section. The concept behind these distributors is to provide the air distribution of a grid while avoiding the mechanical problems associated with grid distributors. A typical pipe grid is shown in Figure 4-5. Air from the main inlet line flows into the main headers and from these into the numerous arms. The arms are fitted with nozzles and the arrangement of arms and nozzles gives complete coverage of the regenerator cross section. Pipe distributors are intended for air only and the spent catalyst enters the regenerator through a separate system.

Properly designed, pipe grids give both good air distribution coupled with mechanical reliability. Problems with this type of air distributor can usually be traced back to either poor design

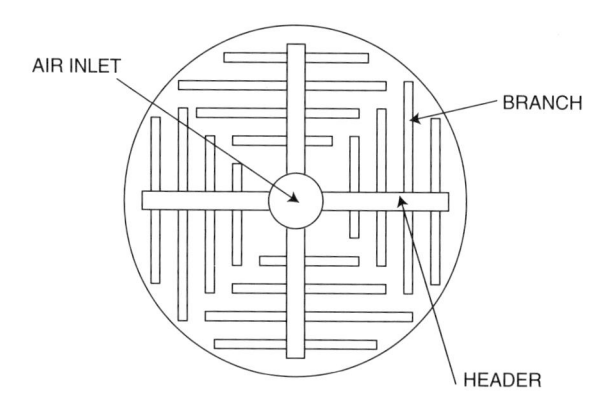

FIGURE 4-5. Pipe grid (plan view).

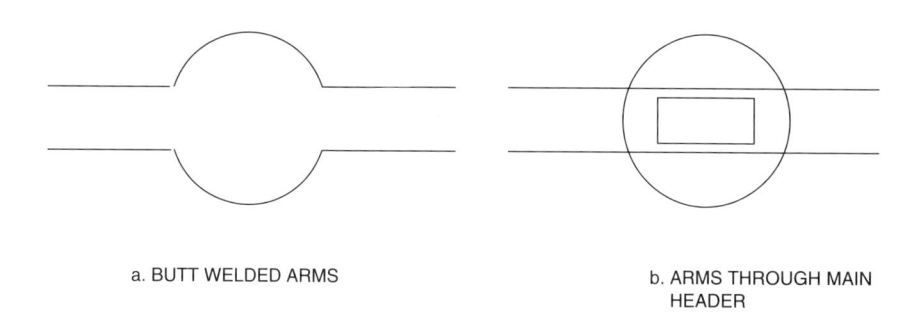

a. BUTT WELDED ARMS

b. ARMS THROUGH MAIN HEADER

FIGURE 4-6. Pipe grid arm joints.

(process or mechanical) or to improper fabrication. Since the headers and arms on a pipe grid are generally self-supporting, proper care must be taken to assure that they are strong enough to withstand the stresses of operation in a hot fluid bed. The distributor arms should not be supported by the welds to the main header (Figure 4-6a) but should pass through the header (Figure 4-6b).

The main headers should be supported as shown in Figure 4-7. Triangular bracing should not be used. A triangularly braced header forms a rigid system. As this system heats up, the lack of flexibility causes stress on the header and the bracing member. This stress often leads to failure of the welds at the points shown in Figure 4-8.

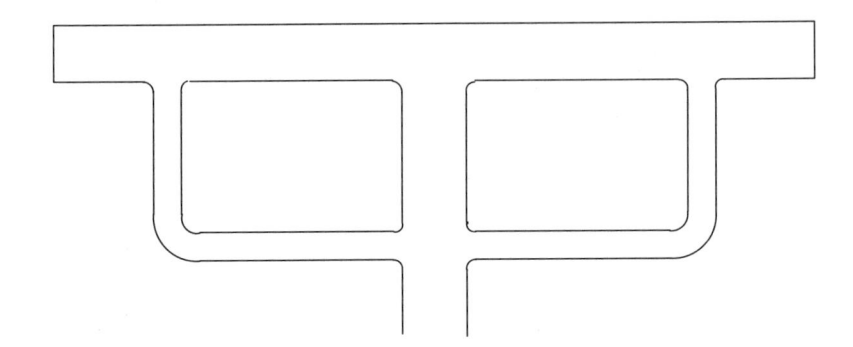

FIGURE 4-7. Pipe grid bracing.

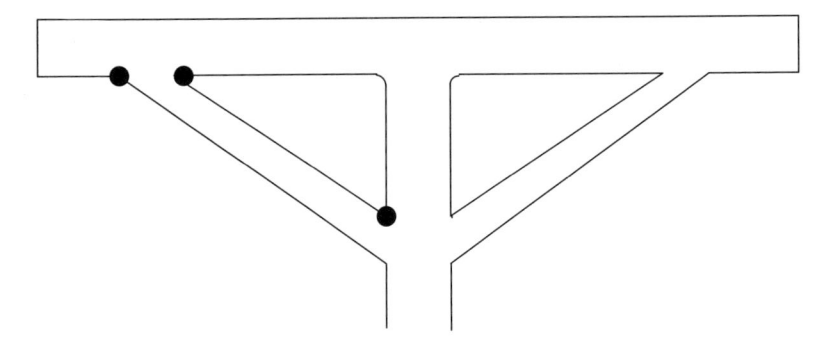

FIGURE 4-8. Air grid cracking on triangular bracing.

The pipe grid should be refractory lined with a 3/4 inch to 1 inch thickness of high density refractory. The refractory can be supported by hexsteel or by S-bar anchors. This lining serves two purposes. First, it protects the metal of the distributor from erosion. Second, it provides enough insulation to the grid to assure that the pipe walls will be at essentially the same temperature at all points. This minimizes the possibility of stress in the pipe from temperature gradients.

Proper placement of the nozzles in the distributor arms is critical to avoid induction of catalyst into the distributor. Any catalyst pulled into the distributor due to an improperly placed nozzle will eventually be blown out of a different nozzle. This will eventually lead to severe erosion of the nozzles and the distributor arms.

JET PENETRATION

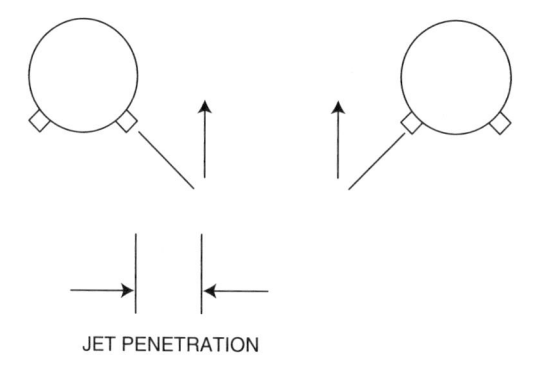

AIR JETS SHOULD NOT CROSS

FIGURE 4-9. Pipe grid arm spacing.

As the air enters the arm from the main header, there will be a zone of low pressure due to the reduction in flow area. To avoid induction of catalyst into the arm, no nozzles should be placed in this low pressure zone.

To minimize interference between air jets, the minimum spacing between nozzles on the same row is 5–6 inches. The spacing between arms should be set to avoid the possibility of jet impingement from the nozzles on adjacent arms. If possible, the arm spacing should be large enough to avoid crossing of the air jets (Figure 4-9). When jets cross, there is generally increased turbulence and thus increased possibilities of erosion. In addition, crossing the air jets can lead to coalescence of the air streams which would degrade the air distribution.

Rings

Rings have been used extensively as vapor distributors in catalytic cracking units. A typical ring distributor is shown in Figure 4-10. Rings offer the advantages of a simple, low cost design with few mechanical problems. In small vessels and in small annular spaces, rings give good distribution. In large regenerator vessels, however, good air distribution can only be achieved through the use of multiple rings or through combinations of rings and other distributors.

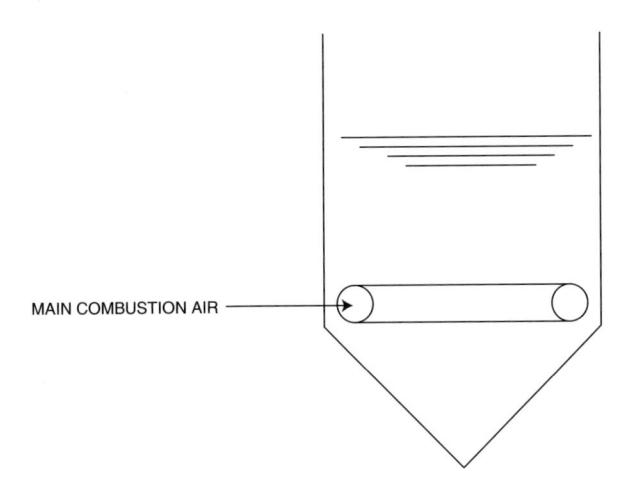

FIGURE 4-10. Air ring.

The design guidelines for rings are essentially the same as for pipe distributors. In addition, rings must generally be supported and these supports must be designed to allow for expansion of the ring as it comes to normal operating temperature.

The primary error found in ring designs is placement of nozzles too close to the entry to the ring. As with the arms on pipe grids, the first nozzle on each side of the ring should be placed sufficiently far from the entrance to avoid any low pressure zone.

Nozzles

Regardless of the type of distributor, if only air passes through the nozzles, dual diameter or Borda tube nozzles (Figure 4-11) should be used. These nozzles consist of a restriction orifice followed by a nozzle tube. The restriction orifice provides the necessary pressure drop to insure good air distribution. Following the restriction orifice, the air jet expands into the tube and the velocity is reduced. Thus, the air leaving the nozzle and entering the catalyst bed is at a moderate velocity. This reduces erosion of the nozzle and minimizes catalyst attrition.

The pressure drop for a Borda tube nozzle can be calculated by the following equation:

$$\Delta P = \frac{K_a V_o^2 \rho}{144(2)g_c}$$

Regenerator Design **141**

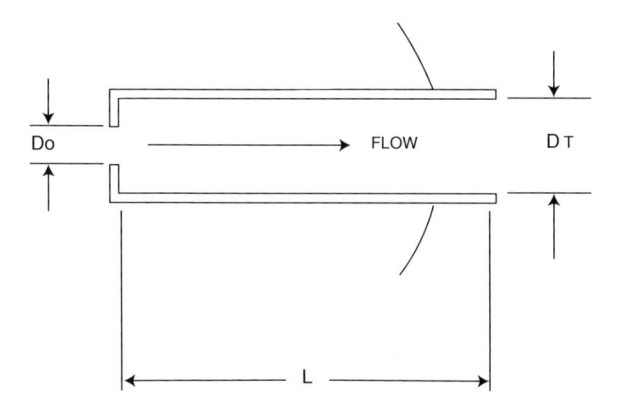

FIGURE 4-11. Borda tube (dual diameter nozzle).

where:

K_a = nozzle coefficient
ΔP = pressure drop, psi
V_o = velocity through nozzle orifice, fps
ρ = gas density at upstream conditions, lb/cu ft
g_c = 32.2

The nozzle coefficient is a function of the diameter ratio of the nozzle orifice (d_o) and the nozzle tube (d_T):

$$K_a = 1.78 - 2\frac{d_o^2}{d_T^2} + 2\frac{d_o^4}{d_T^4}$$

A variation of this design is the 'stick on' nozzle. This nozzle, shown in Figure 4-12, consists of a length of pipe welded on the downstream side of a hole in the distributor. The pressure drop through stick on nozzles can be calculated using the same formula as a Borda tube nozzle. The constant in the equation for K_a should be changed from 1.78 to 1.50 to account for the lower entrance losses for this type of nozzle.

Normally, the length of the nozzle tube should be set at six times the tube diameter. This length will allow the flow in the tube to become fully developed following the nozzle orifice. The internal angle between the orifice and the exit of the nozzle should be 5 degrees or less. For nozzles where d_o approaches d_T, this may require a nozzle length in excess of $6d_T$.

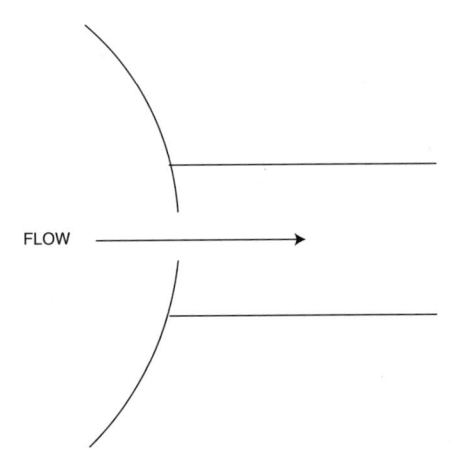

FIGURE 4-12. Stick on nozzle.

If properly designed, nozzles of this type will show only minimal erosion on the nozzle tip. This erosion is due to localized eddies in the catalyst bed caused by expansion of the air leaving the nozzle. This localized erosion can cause refractory damage if the tip of the nozzle is flush with the refractory lining on the distributor. For this reason, the nozzle tube should protrude at least 1 inch beyond the refractory lining.

Air distributors are constructed of various material including stainless steel, low chrome alloys, and even carbon steel. Refractory lined distributors have become the generally accepted design.

Most nozzles are constructed of the same material as the distributor body. Ceramic nozzles have been gaining wider acceptance, however, and there are an increasing number of these nozzles in service. Ceramic nozzles offer improved erosion resistance and are especially useful as inserts in older grids where both catalyst and air flow through the nozzles.

Catalyst Distribution

Since the catalyst contains the fuel (coke) burned in the regenerator, it should be distributed evenly across the regenerator cross section. In many older units, the catalyst entered the regenerator through the grid. This assured excellent catalyst distribution. As mentioned above, however, the catalyst flowing through the grid also accelerated erosion of the grid plate. Newer units, and units with revamped regenerators, generally bring the catalyst into the regenerator separate from the combustion air. In these units catalyst distribution becomes a critical design parameter.

Catalyst can enter the regenerator through standpipes or lift lines. In either case, it can be introduced in the center of the regenerator or near the vessel wall. In the absence of some type of distribution system, the point where the catalyst enters the regenerator will have a high local concentration of carbon.

The air flowing into this high carbon region will not contain sufficient oxygen to completely burn the carbon. Thus, the flue gases leaving this region will be rich in carbon monoxide (Figure 4-13). The air flow to the rest of the bed will often contain oxygen in excess of the amount required. The flue gas from this section of the bed will contain excess oxygen. When these two gas flows mix in the dilute phase or the cyclones, the CO and oxygen will ignite leading to excess afterburning and high local temperatures. Temperature differences across the cyclones of as much as 90°F have been observed in units with poor catalyst distribution.

A frequent remedy suggested for this problem is to adjust the air distribution in the regenerator.[4] This approach is based on the assumption that concentrating the air in either the carbon-rich or

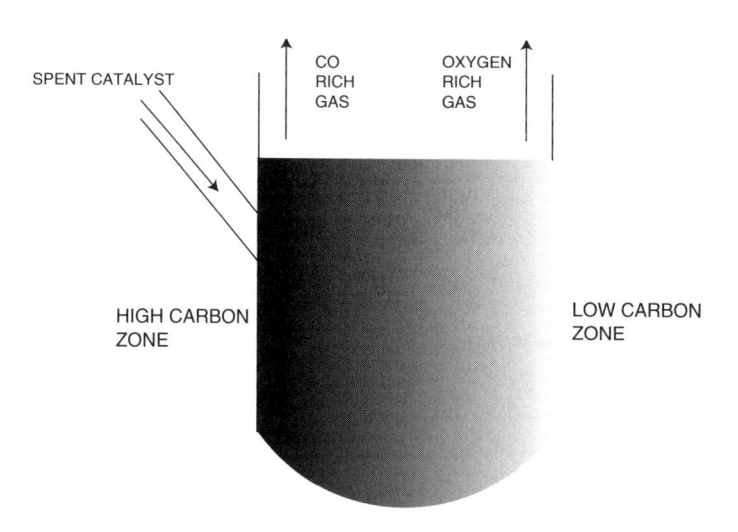

FIGURE 4-13. High carbon zone.

the carbon-poor sections of the bed will promote a more balanced burn.

In practice, these attempts have not always been successful. Even when the distribution of air is shifted to the high carbon zone, it is not possible to burn all of the carbon off in this part of the regenerator (if this were possible, the remainder of the regenerator volume would not be necessary). Thus there will always be carbon on the catalyst in this location.

As long as there is carbon on the catalyst, there will be carbon monoxide in the regenerator. This is due to the following reaction:

$$CO_2 + C \leftrightarrow 2CO$$

Note that oxygen does not play a role in this reaction. Thus, the presence of oxygen in the flue gas has no effect on the equilibrium concentrations of CO_2 and CO.

Given this, shifting a higher percentage of air to the high carbon zone will not prevent afterburning in the dilute phase or the cyclones. In fact, in some units, this shift in air distribution can actually aggravate the problem.

The shift of air to the high carbon zone will result in a higher carbon burn off in this section of the regenerator. This means that the carbon levels in the rest of the regenerator will be lower. Thus, more oxygen may escape from these sections of the bed. This will increase the probability of afterburning in the dilute phase or the cyclones.

Shifting the air preferentially to the sections that are low in carbon can reduce the tendency to afterburn in partial combustion units. In these units, the shift of air away from the high carbon zone decreases the oxygen in the flue gas leaving these zones and thus decreases the oxygen available to react with the carbon monoxide.

In complete combustion units, however, this strategy is less effective. Complete combustion units operate with some excess oxygen in the flue gas. Thus, it is not possible to eliminate the oxygen in the flue gas leaving the bed. If CO escapes from any portion of the bed, it will likely ignite in the dilute phase or in the cyclones.

Since changes in air distribution are not an effective approach to poor catalyst distribution, the catalyst distribution itself must be improved. For units using lift lines to transport the catalyst to the regenerator, this can be done by installing distribution arms or distribution grids. Distribution arms direct the catalyst across a wider section of the regenerator cross section. They do little, however, to improve the distribution of the lift air.

Distribution grids disperse both the catalyst and the air. They are, however, subject to the erosion problems discussed earlier. These can be largely eliminated by using ceramic nozzle inserts in the grid.

Several distribution systems have been used for units where the catalyst enters the regenerator from the spent catalyst standpipe. Ski jump type designs use the momentum of the catalyst to carry it away from the regenerator wall and onto a more fluidized section of the regenerator. These distributors are designed to discharge catalyst above the regenerator bed.

Ski jump type distributors have generally given mixed results. They are more effective on units with small regenerator diameters and well fluidized beds. On units with larger regenerator diameters, the momentum of the catalyst leaving the ski jump is often insufficient to give good distribution across the entire regenerator cross section. In addition, since these distributors discharge into the dilute phase, any volatile hydrocarbons not removed by the stripper will vaporize and burn above the bed. This can lead to localized high dilute phase temperatures.

In many cases it is only necessary to move the catalyst from the wall into the more active portion of the bed. This can be done using standpipe extensions or troughs.

Trough distributors make use of the density differences between the catalyst leaving the standpipe and the catalyst in the regenerator bed. The denser catalyst leaving the standpipe flows along the trough. These distributors will function properly even when fully submerged in the bed. Longer troughs may require fluidization to insure smooth flow.

The UOP high efficiency regenerator solves the catalyst distribution problem by improved mixing. The combustor section of the regenerator operates in a highly turbulent mode of fluidization (fast fluid bed). In addition, the length to diameter ratio of the combustor is large. These two factors combine to produce rapid mixing of the spent catalyst with the catalyst in the combustor and the hot catalyst returned from the upper section of the regenerator. This promotes even distribution of the carbon-containing spent catalyst across the combustor cross section.

Regenerators with internal baffles should be avoided. These baffles are sometimes installed in regenerators to prevent spent catalyst from short circuiting the regenerator bed. Effectively, these baffles create a two-stage regenerator similar to the one shown in Figure 4-1. As with these early two-stage regenerators, afterburning is a common problem in regenerators with internal baffles.

Cyclones

Most regenerators are equipped with two stage cyclones. These are normally mounted inside the regenerator vessel. Stone & Webster/IFP uses externally mounted cyclones on their second stage regenerator vessel. General items on cyclone design and operation will be covered in a later chapter. Items of specific interest in regenerator cyclone systems are the support system (hangers) and dipleg design.

Support Systems

Regenerator cyclone hanger systems must be strong enough to support the weight of the cyclones and at the same time must allow sufficient flexibility to allow for expansion of the cyclone system at regenerator temperatures. In older units, designed for relatively low regenerator temperatures, this was done by supporting the first stage cyclone from an angled strap and the second stage cyclone from the regenerator plenum.[5] As the cyclone system heated up, the angled strap rotated downward to adjust to the changing cyclone dimensions.

As regenerator temperatures increased, especially with conversion to complete CO combustion type regeneration, the required cold position angle of the support strap increased. This resulted in increased lateral stress on the cyclone system and led to mechanical failures.[5]

A number of alternate methods for supporting regenerator cyclones have been developed by licensors and cyclone manufacturers. All of these systems use vertical linkage to support the cyclones, thus eliminating horizontal stresses on the cyclone body. Horizontal movement is accommodated in different ways depending on the support system.[5] All of these systems have been used successfully and the correct choice for any given unit will be a function of unit configuration and cyclone design.

Diplegs

Regenerator cyclones handle large quantities of catalyst. Typically, the catalyst entrainment to the first stage cyclones will equal or exceed the catalyst circulation rate in the unit. Approximately 99% of this entrained catalyst is removed from the flue gas in the first stage. This separated catalyst is returned to the regenerator bed via the first stage cyclone diplegs.

Due to the large flow of catalyst, these diplegs do not generally require check valves. Submerging the diplegs in the regenerator bed provides sufficient sealing to prevent gas from flowing up the diplegs. A splash plate located as shown in Figure 4-14 prevents gas bubbles from the bed from entering the dipleg.

Units with this type of first stage dipleg often experience high catalyst losses during start-up catalyst loading. These losses occur until enough catalyst has been loaded into the unit to raise the regenerator bed level to above the bottom of the diplegs. To minimize these start-up losses, many units have been retrofitted with check valves on the first stage diplegs.

The regenerator second stage diplegs are lightly loaded and are normally equipped with check valves to prevent the backflow of gas up the dipleg. These check valves should be fully shrouded.

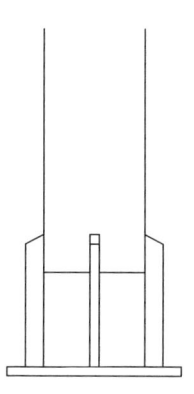

FIGURE 4-14. Primary cyclone displeg splash plate.

Diplegs equipped with check valves should be long enough to insure that the center of the check valve will be at least two feet below the minimum regenerator bed level. Extension of these diplegs to an elevation near the air distributor may lead to failure of the check valves.

Water Sprays

Many older FCCs were equipped with cyclone water sprays. These were located between the first and second stage and were intended to prevent high temperature excursions in the second stage cyclone and the plenum.

These sprays have been largely eliminated for two reasons. First, they were not very effective at preventing high temperatures due to afterburning. Once carbon monoxide burning began in a cyclone, the cooling effect of the sprays was inadequate to control the temperature. Second, poorly designed or damaged spray systems often resulted in direct impingement of liquid water of the inside wall of the cyclone. This resulted in severe damage to the cyclone lining and to the cyclone itself. Regenerator dilute phase sprays have also been largely eliminated for similar reasons.

Plenum Chamber

The outlet tubes from the second stage cyclones are connected to the plenum chamber. In most units, this is an enclosed box located in the top head of the regenerator (Figure 4-15). Plenum chambers of this type have been a frequent source of mechanical problems on FCCs. Even small holes or cracks in the plenum chamber can result in high catalyst losses.

To address these problems, several FCC licensors now offer external plenum chambers. In external plenum designs, the second stage outlet tubes pass through the regenerator head. These tubes then connect to the external plenum. The plenum itself is constructed of cold wall carbon steel pipe. The actual configuration is a function of the unit configuration and the number of regenerator cyclone sets.

External plenums have essentially eliminated the problems associated with other regenerator plenum designs.

Regenerator cyclones, supports, and internal plenums are generally constructed of stainless steel and are designed for con-

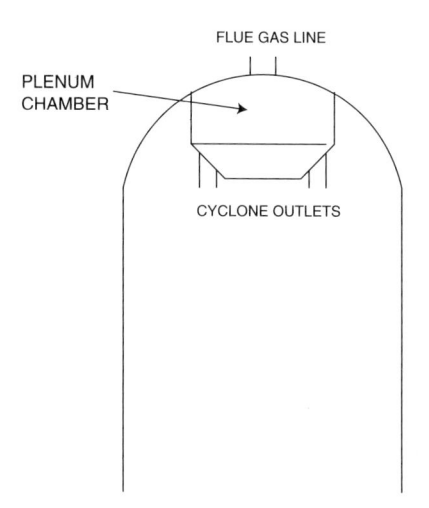

FIGURE 4-15. Internal plenum.

tinuous operation at 1400°F. Some older cyclone systems were made of chrome alloy materials with lower design temperatures.

Air Blower

The air blower supplies the air to burn the coke from the catalyst. A unit operating with complete CO combustion will require 14–15 pounds of air per pound of coke burned. Units operating with partial combustion require 10–12 pounds of air per pound of coke.

Since regenerator pressures are not high, the air blower does not need to develop a high head. Typically, a discharge 10 psi above regenerator pressure will be sufficient to overcome the losses in the air lines and the regenerator.

The primary consideration in air blower design is the ability to deliver a reliable flow of a high volume of air. Both centrifugal and axial air blowers are used in FCC service.

For many years, centrifugal air blowers were the standard for FCCs. These blowers have the advantages of high reliability and a low sensitivity to dirt and other impurities in the ambient air. Their primary disadvantage compared to an axial blower is lower efficiency.

Axial blowers are widely used on new units and on revamps which include the installation of a new air blower. Axial blowers offer higher efficiencies than centrifugal blowers. The high efficiency of axial blowers makes them ideal for systems using a flue gas powered recovery turbine to drive the air blower.

The primary disadvantages of axial blowers are increased sensitivity to dust and a high likelihood of damage if the compressor is surged. Axial blowers are also more expensive than centrifugal blowers.

Carbon Burning Kinetics

In catalyst regeneration, the primary concern is the removal of carbon from the spent catalyst. In most regenerators, this is accomplished in a fluid bed. The velocities in the bed are such that the solid catalyst phase is extremely well-mixed. The gas phase is generally considered to be in plug flow. Based on these assumptions, Luckenbach[6] proposed the following equation for carbon on regenerated catalyst:

$$C_{rc} = \frac{CFT\left(-\ln \frac{O_{2\,out}}{21}\right)}{PKW}$$

where:

C_{rc} = carbon on regenerated catalyst wt.%
C = constant derived from operating data
F = dry air rate, std cu ft/min
T = regenerator temperature, °R
$O_{2\,out}$ = mole % oxygen in dry flue gas
P = regenerator pressure, psia
K = rate constant, sec^{-1}
W = catalyst hold up in regenerator, tons

The kinetic rate constant can be estimated using the following equation:

$$K = 8.286 \times 10^{-4} \exp(0.012\,T)$$

where T is the bed temperature in degrees Fahrenheit.

These equations can be used to estimate the effects of a change in operating parameters such as temperature, pressure, or air rate on catalyst regeneration. To do this, the constant C must be determined using data from well-defined operations. This value will remain valid as long as the equilibrium catalyst properties (ABD, pore volume, surface area) are relatively constant. Once C has been determined, it can be used to calculate the change in carbon that would be expected from a changed regenerator operation.

Regenerator Heat Removal

The earliest FCCs produced more heat from the combustion of coke than was required by the cracking reactions. These units required some means of removing this excess heat. To meet this need, early FCCs were equipped with catalyst coolers. These were essentially shell and tube heat exchangers. The hot catalyst flowed through the tubes and steam was generated on the shell side.

The high velocity catalyst flow through the tubes led to serious erosion problems which limited the reliability of these coolers. Improvements in unit design and operation as well as improved catalysts with better selectivity eliminated the need for regenerator heat removal. Thus, these early coolers fell into disuse.

In 1963, Phillips Petroleum and M.W. Kellogg developed the first design for a resid cracker. At that time, the designers realized that some means of dealing with the excess heat generated by the high coke yields would be necessary. This first HOC used internal regenerator bed coils to remove this excess heat from the regenerator.

Bed coils are horizontal hairpin coils located around the outside circumference of the regenerator bed (Figure 4-16). During operation, boiler feed water was circulated to the coils by means of large circulating pumps. Heat transferred from the bed to the coils generated high pressure steam. This approach to heat removal was successful and this unit is still in operation today.

Bed coils are a viable option provided the heat removal requirement is fixed. The heat transfer to the coils is a function of the regenerator bed conditions and is essentially constant within the boundaries of normal unit operation. In addition, coils cannot be turned off and on during unit operation without suffering damage from thermal shock.

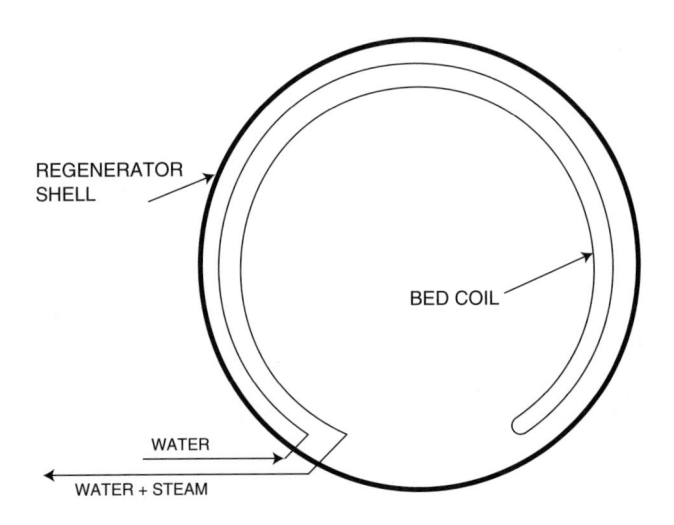

FIGURE 4-16. Regenerator bed coils.

Catalyst coolers offer the flexibility of controlled heat removal. Early attempts to install catalyst coolers on resid FCCs essentially duplicated the shell and tube designs.[5] Not surprisingly, these coolers soon experienced the same problems with tube erosion.

During the development of the RCC process, UOP and Ashland developed a dense phase catalyst cooler in Figure 4-17. In this cooler the catalyst flowed through the shell side of the exchanger. Boiler feed water flowed into the inner pipe of the bayonet tubes and steam and water flowed down the outer annulus. Since both the water in the inlet tube and the steam/water mixture in the annulus are at saturation temperature, no heat is transferred to the inlet tube. Thus, all of the vaporization occurs in the annulus.

The catalyst in the shell is fluidized by air injected through the fluidization lances. Early designs suffered from impingement of the fluidization air on the tubes which led to erosion and tube failure. Later improvements in the tube and air lance layout eliminated these problems.

The first coolers of this type were the flow through design (Figure 4-18a). In these coolers, the rate of heat removal was controlled by varying the catalyst circulation through the heater.

It was soon realized, however, that the heat removal could also be controlled by varying the fluidization air rate and thus the density of the catalyst bed in the cooler. This led to the back mixed design shown in Figure 4-18b.

Regenerator Design **153**

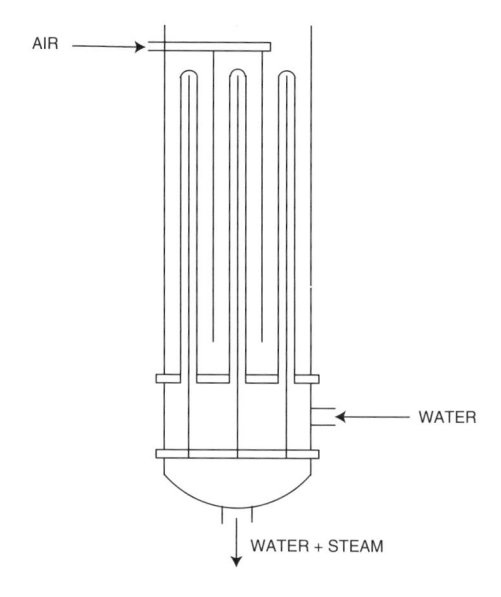

FIGURE 4-17. UOP dense phase catalyst cooler. (Meyres et al., "Improved Resid Processing in FCC Units with Catalyst Coolers," NPRA Annual Meeting Paper AM-94-26. Used by Permission of UOP.)

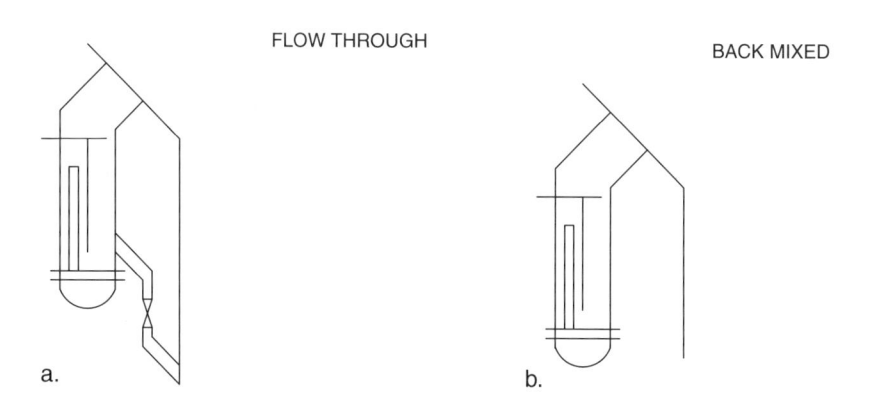

FIGURE 4-18. Flow through and backmix coolers. (Meyres et al., "Improved Resid Processing in FCC Units with Catalyst Coolers," NPRA Annual Meeting Paper AM-94-26. Used by Permission of UOP.)

154   Chapter 4

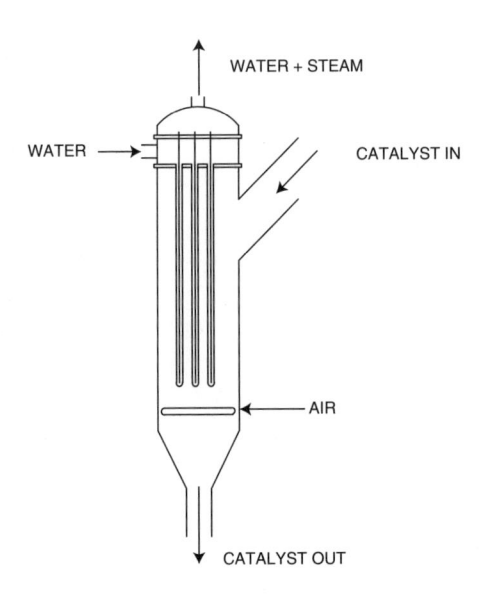

FIGURE 4-19.   M.W. Kellogg dense phase catalyst cooler. (Courtesy M.W. Kellogg.)

UOP catalyst coolers have been widely accepted and are in service on many types of FCC units.

After experimenting with dilute phase coolers, M.W. Kellogg also developed a dense phase cooler design (Figure 4-19). This cooler is similar to the UOP design in that the catalyst flows through the shell side and the water and steam flow through bayonet tubes. The Kellogg design, however, has the tube sheet at the top of the cooler instead of at the bottom. This simplifies the problem of fluidization, but requires an external vent line to direct the fluidization air to the regenerator. This configuration also means that Kellogg designs must be configured as flow through coolers.

Stone and Webster provides a catalyst cooler design where necessary; in many cases, the use of two-stage regeneration eliminates the need for external heat removal. This cooler uses multitube modules submerged in a fluidized catalyst bed. Unlike the UOP and Kellogg coolers, the Stone & Webster cooler has no tube sheet. Instead, each module enters and leaves the cooler separately. The inlets and outlets are equipped with isolation valves that permit the isolation of any module that develops a leak.

As with the Kellogg design, the Stone & Webster coolers use a single air distributor located below the tubes to fluidize the cat-

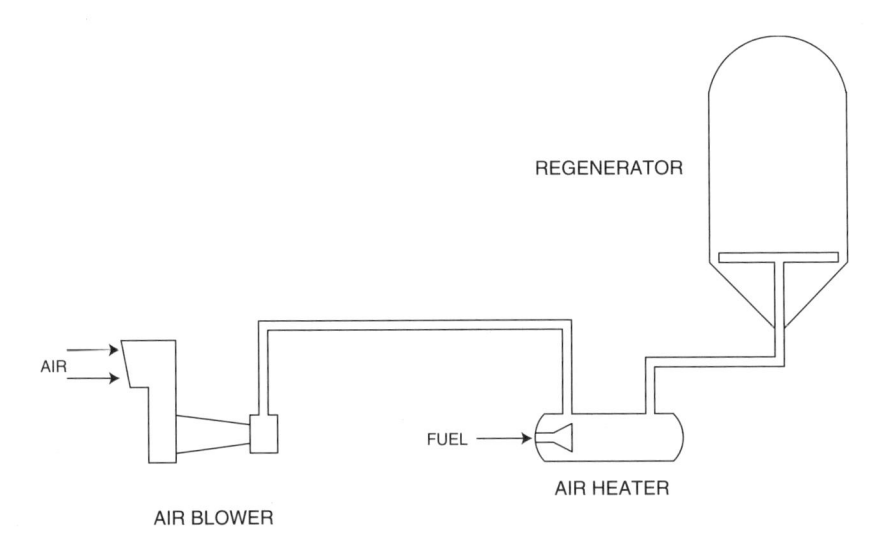

FIGURE 4-20. Air heater.

alyst bed.[7] The air used for fluidization is routed to the regenerator through an external vent line.

Air Heaters

Inline air heaters are used to heat the reactor/regenerator section for start up. They can also be used to dry out the refractory in new or revamped units.

The heater is essentially a gas burner located in the main air line (Figure 4-20). During start-up the burner is fired to heat the air flowing to the regenerator. By adjusting the reactor/regenerator pressure difference and the position of the slide valves, hot air can be directed from the regenerator to the reactor. During start-ups, it is best to heat the reactor to a minimum of 400°F before introducing steam. This prevents steam condensation in the reactor.

Operating Considerations

The regenerator is a common cause of unit limitations as well as the source of many unit upsets. Poor regeneration efficiency, equipment damage, and excessive afterburn are common problems that can often be traced to improper operation. Proper monitoring and control of the regenerator requires attention to several key parameters as well as an understanding of the design limits of the regenerator equipment.

Regeneration Efficiency

The primary consideration in regenerator operation is to provide sufficient air to remove the coke from the catalyst. If this is not done, the carbon level of the regenerator will increase with each pass through the regenerator system. Eventually, this will lead to a carbon snowball.

In a snowball, the level of residual carbon on the catalyst leaving the regenerator is so high that there is little catalytic activity. Instead, the catalyst acts as a heat transfer medium only and the feed entering the riser is thermally cracked. Since thermal cracking is less selective, the coke lay-down increases and the buildup of carbon on the catalyst accelerates or snowballs.

For units in complete CO combustion, maintaining a sufficient air rate to the regenerator generally means holding the flue gas excess oxygen at or above some minimum level. This insures both full regeneration of the catalyst as well as complete combustion of the carbon monoxide. The actual minimum oxygen level required is dependent on the regenerator temperature as well as the design of the regenerator air and catalyst distribution system.

A common design value for the minimum excess oxygen in the flue gas is 2.0 volume percent (vol.%) on a dry basis. Many units operate well at levels as low as 0.5 vol.%, however, while others require excess oxygen levels of 3.0 percent or more.

The need to operate with a high excess oxygen content to avoid afterburn may indicate poor air distribution. Where this is the case, a portion of the air entering the regenerator passes through the bed with little or no contact with the carbon-containing catalyst or with any CO produced by carbon combustion.

This is a common problem in units that use lift air to transport spent catalyst to the regenerator. If the lift line is not equipped with some form of air distributor, the lift air channels through the regenerator bed and provides little in the way of oxygen for regeneration.

The need to maintain a high excess oxygen may also indicate inadequate residence time in the regenerator bed. In this case, the carbon monoxide produced in the bed flows into the dilute phase before it can burn to carbon dioxide. In these units, raising the regenerator bed level will often produce a noticeable improvement in regenerator operation.

A rapidly declining oxygen level in a full burn unit requires immediate corrective action. If the oxygen level is not stabilized, the unit will drop into partial burn. In addition, a rapidly declining oxygen level indicates that the air rate is not sufficient to burn all of the coke produced in the reactor. This could lead to a carbon snowball and a serious regenerator upset.

Controlling the air rate to partial burn regenerators is more complicated. Since there is little or no excess oxygen in the flue gas from these regenerators, oxygen concentration cannot be used to regulate the air rate. Instead, the ratio of carbon monoxide to carbon dioxide (CO/CO_2) must be used.

As the air rate is increased relative to the coke yield, more carbon monoxide is burned to carbon dioxide. Thus, the CO/CO_2 ratio decreases. Conversely, a rising CO/CO_2 ratio indicates that the air rate, relative to the coke yield is falling. Most partial burn units operate with CO/CO_2 ratios of 0.5 to 1.0. Operation at higher values generally does not give good regeneration. Operation at lower values often leads to afterburning in the cyclones.

The key to stable operation of partial burn regenerators is not, however, the absolute value of the CO/CO_2 ratio. Instead, it is maintaining a steady value for this key parameter. Thus, the air rate must be adjusted up or down based on regular flue gas analysis. A rapidly increasing ratio is cause for serious concern. This type of deviation indicates a rapidly increasing carbon buildup on the catalyst and immediate action must be taken to avoid a regenerator upset.

Many units today use online flue gas analyzers to monitor flue gas composition. This allows the operators to keep a constant check on the relevant flue gas parameters and adjust unit operation as necessary. Units with advanced computer control frequently use online analyzers and the control system algorithms to control the air blower operation.

The online analyzers should be calibrated regularly. The flue gas should be sampled and analyzed in the laboratory periodically. This final check will identify any problems with the online analyzer sampling system. Typical problems encountered are air leaks and plugged sampling systems. If not detected, these can lead to poor operation, or possibly to dangerous regenerator upset.

In addition to monitoring the flue gas composition, the color of the regenerated catalyst should be watched closely. This is normally done by taking regenerated catalyst samples two or three times a shift. Successive samples are placed in the cups of a steel muffin tin. When viewed, even small changes in the carbon level on the regenerated catalyst will appear as a difference in color between samples. If used regularly, this procedure will often detect a change in regenerator operation before it becomes apparent from other operating parameters.

Air Rates

Regenerator air rates must fall within maximum and minimum values that are determined by unit design parameters. Usually, the controlling variable is the gas velocity in some part of the regenerator. Areas of concern include the regenerator vessel superficial velocity, the cyclone inlet and outlet velocities, and the air distributor nozzle velocity.

Regenerator superficial velocities normally run between 2.0 and 4.0 feet per second. Lower velocities are possible, but the mixing of catalyst and air is poor and this generally leads to poor regeneration. Higher velocities lead to high rates of catalyst entrainment from the bed into the dilute phase and thus, increased catalyst losses.

The major exception to this criteria is the UOP high efficiency regenerator. In this regenerator, the lower vessel operates as a fast fluid bed. Velocities are much higher and all of the catalyst is entrained from the lower vessel into the upper collection vessel.

High cyclone inlet velocities result in increased erosion of the cyclones and increased catalyst attrition. In addition, high inlet velocities increase the cyclone pressure drop and the level of catalyst in the cyclone dipleg. If the catalyst level in the dipleg rises into the cyclone dust bowl or body, the dipleg is said to be flooded. When this occurs, losses from the cyclone increase significantly.

Air distributor operation is sensitive to low air rates. For the air distributor to operate properly, the pressure drop through the nozzles must be high enough to insure an even flow of air to each nozzle. The nozzle pressure drop must also be high enough to insure that transient localized bed density differences do not affect air distribution.

The minimum nozzle pressure drop is normally set at 30–40% of the bed static head above the distributor.

If the air distributor is operated at an air rate below that needed to give this minimum pressure drop, the air will preferentially flow to nozzles located in localized areas of low density and air distribution will suffer.

In addition, sustained operation at these low air rates may result in catalyst backflowing into the air distributor through some nozzles. This can lead to plugged nozzles, or possibly to severe erosion of the air distributor.

High air rates through the air distributor will produce high nozzle exit velocities. This can lead to increased catalyst attrition and/or accelerated erosion of the distributor nozzles.

It is a good practice to monitor air distributor pressure drop using a dedicated pressure differential indicator. In addition to indicating low air rates, this instrument can also give an early warning of air distributor damage.

Units with plate grids that are not designed for catalyst flow through the grid should monitor the temperature in the space below the grid. This temperature should be approximately equal to the air blower discharge temperature. Higher temperatures indicate that hot catalyst from the bed is leaking through the grid. This could be due to damaged grid seals or grid plates.

In these cases, it is generally best to shut the unit down and make the necessary repairs rather than attempt to continue operations. Continued operation will only aggravate the problem and could lead to extensive damage to the air grid. The catalyst entering the space below the grid will be returned to the bed through the grid holes. This will lead to accelerated erosion of the grid. In addition, the flow of catalyst through the damaged seals or plates will cause further damage to these items.

Afterburning

Afterburning is the combustion of carbon monoxide in the regenerator dilute phase, cyclones, or flue gas lines. Since the catalyst density in these locations is low, there is little mass to absorb the heat generated. Thus, afterburning results in high flue gas temperatures that can damage equipment.

Afterburning is frequently caused by poor catalyst or air distribution in the regenerator. When this is the cause, the afterburning is often limited to one or two cyclones. There may also be a noticeable temperature gradient across the regenerator cross section.

Afterburning that is evenly distributed across the regenerator cross section or the cyclone array usually indicates inadequate gas residence time in the regenerator bed. This can often be corrected by increasing the excess oxygen in the flue gas or by increasing the depth of the bed. Increased promoter additions may also correct this problem.

Carbon Buildup

If the air supplied to the regenerator is not sufficient to burn all of the coke produced in the reactor, there will be a carbon buildup on the regenerated catalyst. If the problem is detected early, it can usually be corrected by increasing the air rate to the regenerator. If this is not possible, then either the feed rate must be reduced, or some other operational change must be made to reduce the formation of coke in the reactor.

If the carbon buildup is severe, however, the problem must be dealt with carefully. Simply increasing the air rate will often lead to a runaway temperature increase as the excess carbon burns away. The temperatures reached in these upsets is generally unknown since it often exceeds the maximum reading on regenerator temperature indicators.

In the event of a serious carbon buildup, the air rates should be increased gradually. The regenerator temperatures must be monitored carefully. At the first sign of a significant temperature increase, the feed rate should be reduced, or if necessary, cut completely. Catalyst circulation to the reactor should be maintained using lift steam if necessary.

Once the production of coke is decreased, the carbon levels can be slowly reduced to normal. Catalyst color should be checked frequently to monitor the level of carbon removal. High temperature excursions may occur as the excess carbon is burned. It is usually best to ride out these excursions by maintaining a controlled air rate and maximizing catalyst circulation to the reactor. If catalyst circulation is halted, the regenerator temperatures will increase due to the loss of heat removal by the circulating catalyst. If the air blower is tripped, the hot catalyst in the regenerator bed will defluidize (slump). The defluidized bed will lose heat very slowly. A large bed may require several days or even weeks to cool.

References

[1] U.S. Patent No. 3,909,392.

[2] *FCC Unit Design, Operation and Troubleshooting,* Refining Process Services Seminar (1993).

[3] Pfeiffer, R.W., et al., U.S. Patent No. 3,563,911.

[4] Cabrerra, C.A., Lacijan, L.A., and Lunda, M.O., "Mechanical Considerations in FCC Design," Ketjen Catalysts Symposium (1984).

[5] Sloan, H.D., Wilson, J.W., and Glasgow, P.E., "Revamp of FCC Units to Process Resid," Ketjen Catalysts Symposium (1984).

[6] Luckenbach, E.C., "How to Update a Catalytic Cracking Unit," *Chemical Engineering Progress,* February (1979).

[7] Letzsch, W.S., Dharia, D.J., Wallendorf, W.H., and Ross, J.L., "FCC Modifications and Their Impact on Yields and Economics," NPRA Annual Meeting Paper AM-96-44 (1996).

ॐ5
Flue Gas Systems

Flue Gas Flows and Properties

Regenerator flue gases contain carbon dioxide, nitrogen, oxygen and water. If the regenerator is not operating in complete CO combustion, the flue gas will also contain carbon monoxide. In addition to these gases, the flue gas contains any catalyst particles not captured by the regenerator cyclones. The flue gas also contains significant energy in the form of elevated temperature, pressure, and possibly unburned carbon monoxide. Typical flue gas compositions (mole percent dry basis) are:

	Complete Combustion	Partial Combustion
Oxygen	2.00	0.00
Carbon monoxide	0.05	6.00
Carbon dioxide	16.00	12.00
Nitrogen	82.00	82.00

In addition, the flue gas will contain sulfur oxides (SO_x). The concentration of SO_x in the flue gas will depend on the sulfur content of the feed. Approximately, 10% of the sulfur from unhydrotreated feeds will go to the coke. This sulfur will appear as SO_x in the flue gas. The percentage of feed sulfur in the coke will be higher for hydrotreated feeds. Due to the low overall feed sulfur content, however, the total SO_x in the flue gas will be less. Since resid feeds generally produce higher yields of coke, a larger percentage of the feed sulfur will appear in the flue gas. Typical values for resid feeds were given in Chapter 2.

The volume of the flue gas flow can be quite large. A nominal 30,000 BPSD FCCU yielding 5.0 wt.% coke from a vacuum gas oil feed will produce approximately 63,500 standard cubic feet of flue gas per minute. The purpose of the flue gas system is to control this flow and to extract an economic amount of energy from this gas stream.

Flue Gas Systems

The simplest flue gas system is shown in Figure 5-1. Flue gas leaving the regenerator passes through a slide valve which is used to control regenerator pressure. An orifice chamber downstream of the slide valve reduces the pressure further, and the hot flue gas then flows into the stack and to the atmosphere.

While this system is simple, it has several negative aspects. First, there is considerable energy lost in venting the flue gas to the atmosphere. Second, there is no attempt to remove catalyst particulates or other undesirable components of the flue gas.

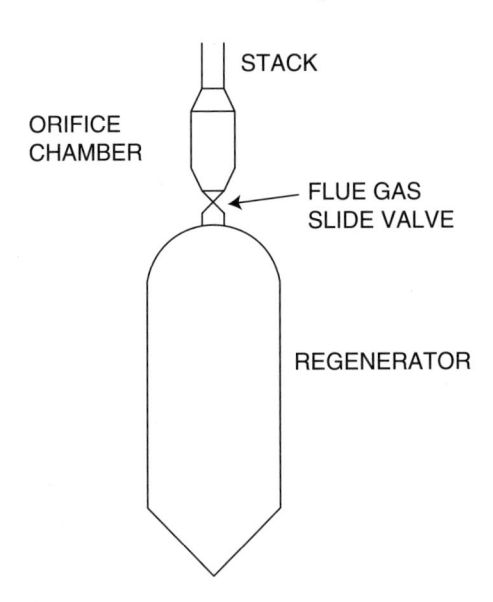

FIGURE 5-1. Simple flue gas system.

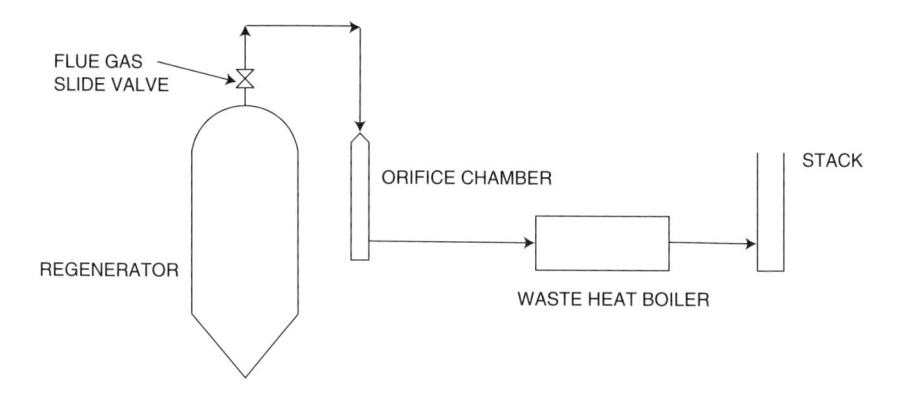

FIGURE 5-2. Flue gas system with waste heat boiler.

A more typical flue gas system is shown in Figure 5-2. The hot flue gas flows through the slide valve and orifice chamber as before. From this point, the flue gas flows to a flue gas cooler. Here, the thermal energy in the flue gas is removed and used to generate steam.

In units where the carbon monoxide is present in the flue gas, the flue gas cooler is replaced by a CO boiler. Here, the flue gas is mixed with air and the carbon monoxide is burned to CO_2. Supplemental fuel is used to assure that the CO is completely oxidized. Steam is generated in the boiler section.

Figure 5-3 shows a very sophisticated flue gas system. Flue gas from the regenerator passes through a butterfly valve which is used to control the regenerator pressure. The flue gas then passes through a third stage separator to remove catalyst fines and a power recovery turboexpander where the pressure energy is converted to mechanical energy. The energy recovered in the turboexpander is used to drive the air blower.

Following the turboexpander, the flue gas passes through a cooler and then into a flue gas desulfurization unit (desox). The flue gas from this unit flows through the stack to the atmosphere.

The optimum flue gas system will depend on local conditions, regulations, and the value of recovered energy. The trend, however, has been toward increased sophistication and increased levels of downstream cleanup of the flue gas. In the near future, it is likely that most FCCs will be equipped with at least some of the features shown in Figure 5-3.

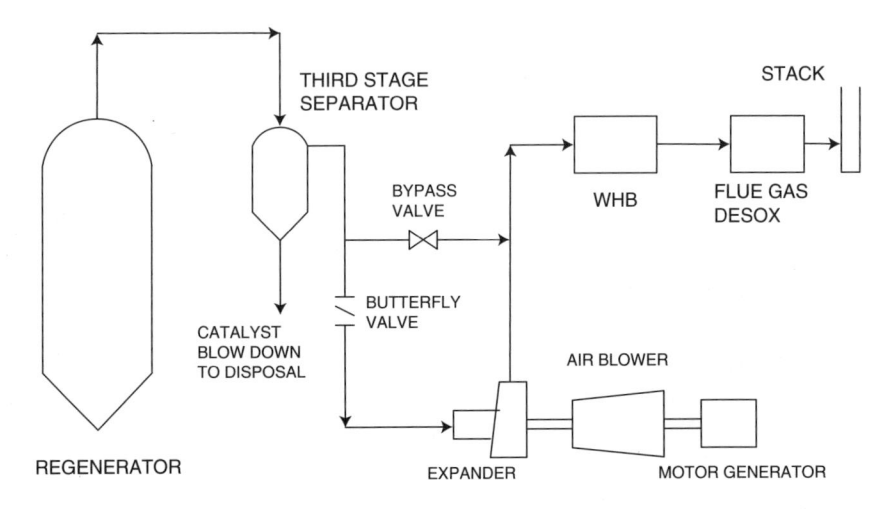

FIGURE 5-3. Flue gas system with power recovery.

Flue Gas Control Valves

The flue gas control valve(s) regulate the flow of the flue gas from the regenerator and thus, the regenerator pressure. Usually, these valves are double disk slide valves and often, there are two valves installed in series. The control system is usually set up so that each disk can be placed on automatic. This should not be done as it will often lead to cyclic operation of the valves as each disk tries to catch up to the others. This can be particularly troublesome when the valve has pneumatic operators which are slower and have less precise positioning than hydraulic operators.

Normally, one disk is placed on automatic while the others are positioned manually to assure that the automatically controlled disk remains in control. It is good practice to periodically change the disk which is on automatic control.

Flue gas systems with turboexpanders often use butterfly valves to control regenerator pressure. These valves result in less valve pressure drop and thus, higher inlet pressures to the turboexpander. Since the purpose of the turboexpander is to convert pressure energy to mechanical energy, higher inlet pressures result in higher energy recovery.

Flue gas control valves are generally designed to have a minimum opening. This is intended to prevent full isolation of the regenerator vessel. In slide valve systems, the minimum opening normally consists of cutouts in the disks. When the valve is fully closed, these cutouts provide an open area to prevent overpressuring of the regenerator vessel.

Butterfly valves used in flue gas service are generally equipped with stops which prevent full closure. These serve the same purpose as the slide valve cutouts. Mechanical stops which block the disk are preferred over software stops. Software stops can be changed through reprogramming of the control system and thus, do not provide as much protection.

Orifice Chamber

The orifice chamber is used to reduce the pressure drop across the regenerator flue gas slide valve. Without an orifice chamber, the flue gas velocity through the slide valve would be high and would generate considerable noise. A typical orifice chamber contains several perforated plates (Figure 5-4). The pressure drop across each plate can be calculated by the following formula:[1]

$$\Delta P = \frac{V^2 \rho \left[1 - \left(A_f/A_p\right)^2\right]}{2 g_c \, C^2 Y^2}$$

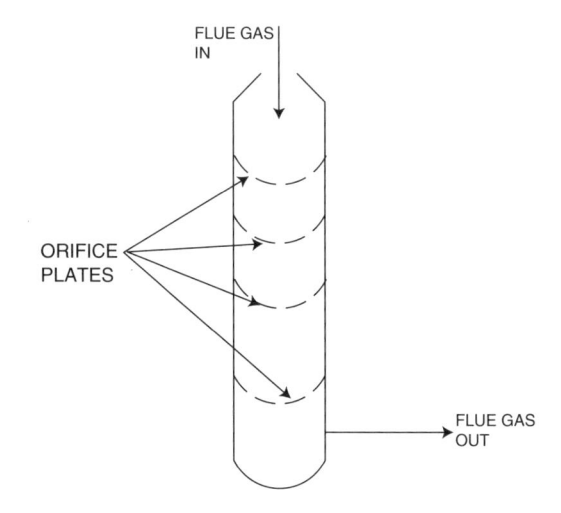

FIGURE 5-4. Orifice chamber.

where:

ΔP = pressure drop through plate, lb/ft^2
V = velocity through orifices, ft/sec
ρ = gas density at upstream conditions, lb/ft^3
A_f = orifice plate free area, ft.
A_p = orifice plate area, ft.
g_c = 32.2
C = discharge coefficient
Y = compressibility factor

C can be estimated from the following equation:

$$C = \frac{0.586339 + 0.658563(t/D) - 0.34458(t/D)^2 + 0.057288(t/D)^3}{(P/D)^{0.10}}$$

where

t = plate thickness
D = hole diameter
P = hole pitch

This equation is valid for t/D between 0.4 and 2.4 and for hole Reynolds numbers between 4,000 and 20,000. These conditions will cover most orifice chamber designs. Figure 5-37 in the cited reference can be used to estimate C for designs that fall outside of this range.[1]

The overall pressure drop for the orifice chamber is set to give a reasonable slide valve pressure drop. Once the overall ΔP is set, the number of plates and the free area for each plate are set to provide a smooth pressure drop across the chamber.

Excessive velocities across individual plates should be avoided since this can lead to vibration of the plates and damage to the orifice chamber. If necessary, additional plates should be added.

Third Stage Separators

Third stage catalyst separators can be used to reduce the particulate emissions from the FCCU regenerator and/or to protect a downstream turboexpander. Several designs are available.

Shell Separator

The most widely used design is the Shell separator in Figure 5-5. This separator was developed by Shell to protect turboexpanders from catalyst particles in the flue gas.

The separator consists of a vessel which contains numerous swirl tube separators. These separators are small axial flow cyclones (Figure 5-6). Flue gas entering the separator tube passes through the swirl vanes which impart a spinning motion to the gas

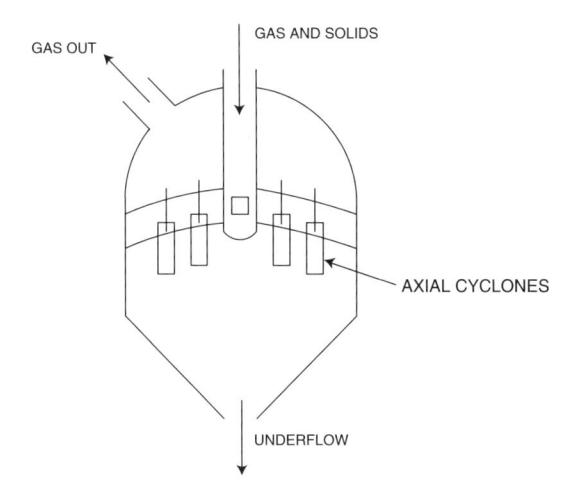

FIGURE 5-5. Shell third stage separator.

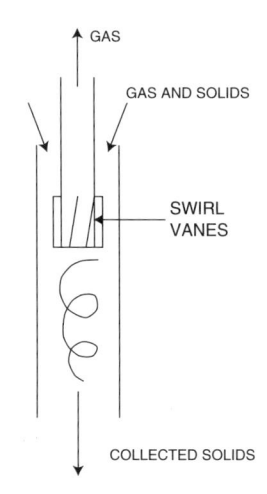

FIGURE 5-6. Axial cyclone used in shell separator.

flow. The resulting forces move the catalyst particles to the tube wall where they are separated from the gas steam.

Separated particles fall through the bottom of the tubes and collect in the conical bottom of the separator vessel. A controlled underflow of flue gas carries these particles from the separator to final disposal.

Early Shell separator designs use closed bottom separators. The catalyst drained through small holes in the base of the tube. The closed tube designs were easily overloaded by unexpected catalyst losses. When this occurred, the catalyst passed through the separator and damaged the expander. Later designs using open bottom tube separators are less likely to be overloaded by sudden catalyst losses.

Polutrol Separator

Polutrol separators use a patented arrangement of small horizontal cyclones inside a specially designed containment vessel. The placement of the cyclones and the design of the vessel make this separator very tolerant of sudden increases in flue gas dust loading.

Collected dust can be removed by a continuous underflow or can be allowed to collect in the bottom of the separator and removed by periodic gravity drainage.

As the name Polutrol implies, this separator was originally developed to reduce particulate emissions. It has, however, been proven to provide very effective protection to flue gas turboexpanders.

Mobil/M.W. Kellogg Separator

M.W. Kellogg licenses a third stage separator incorporating improved cyclone designs developed by Mobil. Kellogg and Mobil claim increased separation efficiencies over previous separator designs.

Other Separator Designs

Other third stage separators based on more conventional cyclone designs are available from G.E., Emtrol, and van Tongeren. All of these separators will give good performance. Typically, the flue gas solids content can be reduced to 100–150 mg/Nm^3.

Turboexpanders

Turboexpanders, also known as power recovery turbines, convert a portion of the pressure energy in the flue gas into mechanical energy. This recovered energy is generally used to drive the regenerator air blower, but may also be used to do other work such as driving an electric generator.

Early attempts at recovering pressure energy from FCC flue gas using commercially available turboexpanders were noted for their failure.[2] The primary problem was rapid erosion of the turbine blades by the catalyst present in the flue gas. These early experiments made two things obvious. First, a reliable means of removing catalyst fines from hot flue gas was necessary. Second, the turboexpanders used for FCC power recovery should be designed specifically for that service.

The first successful FCC power recovery systems were developed by Shell. These systems use the Shell third stage separator described earlier coupled with single stage expanders designed for FCC flue gas service.

This development was so successful that Shell-licensed power recovery systems dominated FCC applications. The two major licensors of FCC technology of the time—UOP and Kellogg—signed on as agents for the Shell system.

While the Shell system is still widely used in FCC power recovery applications, successful competitors have emerged. These alternative systems are generally based on a third stage separator designed by a cyclone vendor coupled with a purpose built expander.

There have also been some notable failures. Several installations using multistage expanders resulted in early rotor failures. These failures led many to believe that multistage expanders were not viable in FCC flue gas service.

In actuality, the problem was more likely the result of poor turbine design. Many of these multistage designs used a rotor supported on both ends by bearings (between bearings shaft support) and radial gas entry. Most single stage expanders used an overhung rotor with axial flue gas entry. Two-stage expanders using an overhung rotor and axial flue gas entry have proven to be reliable in FCC service. At least one expander of this type has been in service for over 10 years. Two-stage designs of this type offer higher efficiency, especially on FCC units with regenerator pressures above 30 psig.[2]

The justification for a turboexpander installation is in the recovered energy. Thus, high energy costs increase the economic incentive for this equipment. In addition, units with higher operating pressures have more available energy, and thus higher possible returns. Generally, the potential returns from small low pressure units does not justify power recovery. Larger units, especially in locations with high energy costs, will often see very high returns on the capital invested in a power recovery system.

The power recovered can be calculated by the following equation:[2]

$$HP = \left(\frac{WRT_1}{33,000}\right)\left(\frac{k}{k-1}\right)\eta_o\left[1-\left(\frac{P_2}{P_1}\right)^{\left(\frac{k-1}{k}\right)}\right]$$

where:

HP = horsepower developed by turbine
W = gas flow, lb/min
R = gas constant = 1545.3
T_1 = inlet temperature, °R
P_1 = inlet pressure, psia
P_2 = outlet pressure, psia
k = specific heat ratio of flue gas
η_O = overall expander efficiency

Typical expander efficiencies are 80% for a single-stage machine and 82–85% for multistage expanders.

The installation of a turboexpander alters the operating economics of the FCC. Turboexpanders are constant volume machines. Thus, they function as an orifice in the flue gas line. As the flue gas rate changes with changes in FCC operation, the expander inlet valve will adjust the pressure at the expander inlet to maintain a constant inlet gas volume.

This constant volume operation means that turboexpanders are most effective when operated at their design flow rates and inlet pressures. At low flow rates, the flue gas is throttled by the expander inlet valve. This results in a low expander inlet pressure and thus, a decrease in available power. Higher flue gas rates will result in increased power recovery up to the point where the expander bypass must open to control regenerator pressure. From this point on, increases in flue gas rate will have no effect on recovered power.

The power recovery turbine will provide a maximum percentage of the air blower power requirement over a fairly narrow range of flue gas rates and regenerator pressures. Operation outside of this envelope will increase the amount of external energy required by the FCC.

Most FCC operations are geared toward minimizing the air blower horsepower requirements by minimizing the air rate and minimizing the regenerator pressure. This practice, when applied to a unit equipped with a power recovery turbine, may result in less than optimum operation. In these cases it would be better to operate at a higher air rate and/or a higher regenerator pressure so as to minimize the amount of outside energy required to drive the air blower.

The most common problem associated with turboexpander installations is blade erosion or other damage due to high particulate levels in the flue gas. These are generally the result of poor third stage separator performance or high catalyst losses during unit upsets.

Sudden catalyst losses are undesirable on any FCC, but they assume increased significance on units equipped with turboexpanders. A large catalyst loss from the regenerator can completely overwhelm the third stage separator. When this occurs, catalyst flows directly to the turboexpander where it can cause significant damage to the turbine blades.

Turboexpanders are also subject to a gradual buildup of catalyst deposits on the rotor blades. These deposits are made up of very fine catalyst particles that fuse together on impact with the blades. Over time, this buildup can affect the balance of the turbine rotor and lead to increased vibration.

The deposits can be removed by introducing walnut hulls or other suitable material into the expander inlet line. The deposits can also be removed by thermal cycling of the expander. This is done by opening the expander bypass and allowing the expander to cool. The flue gas flow is then redirected to the expander to reheat the turbine. Several refineries practice thermal cycling to clean expander turbine blades. Expander vendors, however, do not recommend this procedure.

The addition of a turboexpander to the FCC flue gas system increases the maintenance cost for the FCC. In addition, it will have a negative impact on unit reliability.[3] On average, the turboexpander can be expected to result in one to two days of lost operation per year.

Water Sprays

Flue gas systems equipped with power recovery turbines should also be equipped with water sprays to protect the turbine from high flue gas temperatures. These sprays should use steam atomized nozzles and the control system should be set up to insure that the full flow of atomizing steam is established before water is introduced to the nozzles. Mechanical atomization nozzles should be avoided.

Poorly atomized water will not vaporize quickly in the flue gas line. In some cases, poor water atomization has led to serious refractory damage in flue gas lines.

Flue Gas Coolers

The flue gas also contains considerable thermal energy. This can be recovered and used to generate steam. Early flue gas coolers were essentially large shell and tube exchangers with the flue gas on the tube side and water and steam on the shell side.

These coolers were often placed upstream of the regenerator slide valves and were intended for operation at regenerator pressure.

Modern flue gas coolers are box coolers with the flue gas in the cooler box and water/steam in the tubes. These coolers are in many ways similar to the convection section in a furnace. There are, however, some critical differences between flue gas coolers and furnace convection sections that should be considered.

The first of these is the presence of catalyst in FCC flue gas. Even with sophisticated third stage separators, the catalyst content of FCC flue gas will be on the order of 100 mg/Nm^3. In installations without third stage separators, the loading can be as high as 500 mg/Nm^3. This catalyst is very fine, particle diameters are generally 20 microns or less. These particles have a tendency to collect on surfaces in the flue gas system, and the tubes in the flue gas coolers are particularly prone to fouling.

For this reason, the use of extended surface tubes should be approached with caution. Tubes with closely spaced fins are especially risky as these surfaces will become fouled with catalyst fines in a matter of days. Soot blowers are often completely ineffective in removing the catalyst deposits between the fins.

The safest option is to use smooth tubes. These will foul less, and the deposits will be effectively removed by soot blowers. Tubes with extended surfaces have been used successfully in several FCCs,[3] however, and these will increase the heat recovery. If extended surface tubes are to be used, the supplier of the cooler should provide a list of successful applications in FCC flue gas service. The performance of these installations should be checked to confirm their satisfactory operation.

The flue gas cooler should be equipped with soot blowers to remove catalyst deposits from the tubes. Retractable soot blowers are the most effective. Rotary soot blowers are generally less effective, but have been used successfully in the back end of large coolers.[3]

Cold end corrosion has also been a problem in flue gas coolers. This occurs when the sulfur oxides and water in the flue gas condense as sulfuric acid. The temperature at which this occurs is known as the acid dew point. This can be estimated from the concentrations of water and sulfur oxides in the flue gas.

It is not sufficient that the bulk flue gas temperatures be above the acid dew point. Condensation can occur on cold tube surfaces. This is a common problem where the coldest tube bank in the cooler is used to preheat boiler feed water. As a general rule, the flue gas temperature leaving the cooler should not be less than 50°F above the acid dew point. If the back end of the flue gas cooler is used to preheat boiler feed water, the tube metallurgy should be resistant to acid attack.

Electrostatic Precipitators

Many older FCCs were equipped with electrostatic precipitators (ESPs) to recover catalyst lost from the regenerator cyclones. At the time, this was a cost containment and not a pollution control measure. The catalysts of the day were easily broken up into smaller particles and the cyclone separators available were less efficient than those used today.

Today, ESPs are used to reduce the flue gas solids content in order to meet environmental regulations. A well-designed and well-operated system can reduce the flue gas particulate loading to approximately 20 mg/Nm^3. Even lower particulate loadings are possible provided the refiner is prepared to meet the cost of a large precipitator.

ESPs are widely used to control solids in the flue gases from power boilers and other direct-fired heaters. FCC flue gas applications, however, represent only a small fraction of the total number of ESPs in service. Thus, the number of vendors with FCC experience is limited.[4] The operating characteristics of an FCC are significantly different than those of a direct-fired heater and these differences will have an effect on precipitator design and performance. When considering an ESP for use on an FCC, the vendor's experience with this type of application should be considered. Experience with other precipitator applications is not directly transferable to FCC service.

ESPs collect catalyst fines by inducing an electrical charge on the particle followed by collection of the charged particle on an oppositely charged collection plate. The charged particles are then removed from the plate, collected, and sent to disposal.

The particles are charged through the formation of a nonuniform electrical field between the discharge electrode and the collection electrode.[4] This nonuniform field results in a flow of electrons from one electrode to the other. Some of these electrons strike and thus, ionize flue gas molecules. These ionized molecules also flow to the collection electrode. The combined flow of electrons and gas molecules makes up the corona current. The corona current charges the catalyst particles in the gas stream, and these charged particles are attracted to and collected by the collection electrode.

Larger particles absorb more electrons than smaller particles and are thus more easily collected. Thus, these larger particles are collected toward the front of the precipitator while the smaller particles are collected at the back. This means that the separation process increases in difficulty as the flue gas moves through the precipitator. In addition to particle size, particle composition and flue gas composition also affect particle charging and thus, collection efficiency.[4]

Particle composition affects resistivity, and particles with higher resistivity are more difficult to charge and collect. For FCC catalyst, the key variable is the alumina content of the catalyst. Catalyst with a higher alumina content will have a higher resistivity.

Flue gas composition and temperature affect the charge strength that can be applied to the precipitator. Higher temperatures and increased water content both act to improve precipitator performance. The addition of ammonia to the flue gas can also improve charge strength and thus, precipitator performance.

The charged particles are attracted to the collection plate. The force acting on the particle is given by:[4]

$$F_e = EQ$$

where:

F_e = force on the particle
E = electric field
Q = particle charge

Particle collection requires current flow, and this is affected by both the catalyst resistivity and the thickness of the catalyst layer on the collection plate. This is expressed by:[4]

$$I = \frac{E_a A}{\rho T}$$

where:

I = current flow, amperes
E_a = applied voltage
A = collecting area, cm^2
ρ = catalyst resistivity, ohm-cm
T = thickness of catalyst layer on plate, cm

As this equation shows, increasing catalyst resistivity or an increase in the thickness of the catalyst layer on the plate will reduce current flow. This, in turn, will decrease particle separation.

Once the particles have been collected, they must be removed from the collection plates. As discussed above, the buildup of catalyst on the collection electrode reduces the flow of current and thus, the rate of particle collection. In addition, a layer of catalyst dust will also collect on the discharge electrode. This will increase the electrode diameter, increasing the voltage level required to establish corona flow.[4]

Particles are removed from the electrodes by periodic rapping. Proper rapping is essential to maintaining precipitator performance. Inadequate rapping will lead to poor collection efficiency due to the buildup of catalyst on the electrodes. Too frequent rapping, however, will lead to excessive re-entrainment as the thin dust layer will disperse as it falls from the plates.

The dust that has been dislodged from the electrodes falls into the dust collection hoppers. This dust must be removed and sent to disposal. If the dust is not removed, the level in the hoppers will eventually build up until catalyst is re-entrained into the gas stream. At extreme levels, the catalyst in the hoppers can contact the bottom of the high voltage frame. This will result in short-circuiting of the precipitator. The hoppers are not intended for storage of catalyst fines and they should be drained frequently.[4]

Catalyst can be removed by either gravity drainage or by pneumatic transport. Gravity drainage is the simplest system and is preferred for precipitators operating at or near atmospheric pressure. The dust is dropped directly from the hoppers into containers which are carried away to disposal.[4]

Pneumatic transport systems are more complex. They are preferred where the precipitator operates at significant positive pressure or where the fines must be transported over some distance. Since catalyst fines are cohesive by nature, pneumatic systems are subject to plugging.

The hoppers themselves must be designed for easy drainage. The hopper walls should have an angle of no less than 60 degrees from horizontal. The hoppers should be insulated and heated.[4] The catalyst falling into the hoppers will contain entrained flue gas which contains considerable water. If the hoppers are not kept hot, this water will condense along with any sulfur oxides in the flue gas. This will lead to problems with hopper drainage as well as corrosion of the hoppers.

Each hopper should be equipped with a level instrument to warn of a high hopper level and a temperature indicator to monitor the performance of the heating element.[4]

Most precipitators used in FCC service use weighted wire discharge electrodes. These are subject to failure due to vibration or excessive arcing. Recently, the major ESP manufacturers have developed rigid discharge electrode systems. These are more durable than weighted wire systems.

The installation of an ESP adds a new class of hazards to the FCCU. These risks are different than those normally encountered by FCC operations and maintenance personnel, and they carry a higher potential for fatality. Schiller estimates that 98% of the accidents involving an ESP result in death.[4]

The ESP is essentially a very large capacitor. A precipitator will retain sufficient charge to cause serious injury or death for several hours after the power is turned off. The precipitator must be fully grounded before entry or before contact with any high voltage component. Key interlock systems that prevent entry or access to the high voltage components unless the system is de-energized and grounded should be incorporated into all access points. These systems should never be bypassed or defeated.[4]

Fires and explosions have occurred in electrostatic precipitators. These have generally been due to the ignition of a combustible air/gas mixture by sparking.[4]

Some sparking in an ESP is normal, and in fact desirable. When an explosive mixture is present, however, sparking serves as an ideal source of ignition. Some conditions that can result in an explosive mixture are:

1. *Unit start-up.* During unit start-up, combustibles will sometimes build up in the flue gas system. The source of these combustibles may be the start-up air heater, partially burned torch oil or unvaporized feed. Combined with the high oxygen levels in the flue gas that are common during start-up, these gases can form an explosive mixture in the ESP.

    If possible, the ESP should not be energized during start-up. If this is not possible due to local emissions regulations, the voltage should be reduced to avoid sparking.

2. *Unit upsets.* FCC upsets that result in hydrocarbons entering the regenerator can lead to oil vapors in the flue gas system and the ESP. In addition, high flue gas CO levels and/or high flue gas oxygen levels could also lead to the formation of an explosive mixture in the ESP. To avoid a possible fire or explosion, the ESP should be tripped out on low slide valve differentials or low reactor temperatures.[4]

SO$_x$ Removal

With increasing environmental regulation, it has become necessary to remove sulfur oxides in the flue gas from some FCCUs. Generally, this is done with a flue gas scrubber.

The most widely used system is the Exxon flue gas scrubber. This technology was developed specifically for use on cat crackers. The scrubber uses a Venturi to contact the flue gas with caustic scrubbing solution. The caustic reacts with the sulfur oxides to form sulfates.[5]

The solution is nonregenerable and fresh caustic must be added to maintain the solution strength. Spent caustic must be neutralized and treated before disposal.

Other scrubbing technologies are less widely used on FCCUs. Statoil's RFCC located at Monkstad, Norway uses an ABB Flakt sea water scrubber to remove SO$_x$. Sea water contains natural buffers and is capable of absorbing a considerable quantity of sulfur oxides. In the Statoil installation, sulfide-containing sea water is diluted with sea water used in the refinery cooling system. This diluted effluent is returned to the ocean.[6]

The Mitsui process uses activated coke to remove flue gas SO$_x$. Flue gas and coke are contacted in a moving bed. The activated coke absorbs the sulfur oxides in the flue gas.[7] The process is regenerable and the absorbed SO$_x$ can be recovered as elemental sulfur or as sulfuric acid.

The activated coke process can also absorb nitrogen oxides (NO$_x$). The absorbed NO$_x$ is then reacted with ammonia to produce nitrogen and water. The Mitsui process has been sucessfully installed in Japan on a 30,000 BPSD FCCU.[7]

The Belco EDV process has been installed on several FCCUs.[8] The EDV process can remove a variety of contaminants from the flue gas.

NO$_x$ Removal

Nitrogen oxides are a major contributor to photochemical smog. These pollutants are formed in any high temperature combustion process through fixation of nitrogen present in the combustion air. In addition, basic nitrogen in the FCC feed will react with the acid sites on the catalyst. This nitrogen will be removed in the regenerator and a portion will be converted to nitrogen oxides.

NO_x control on FCC regenerator flue gas is a relatively recent practice. Nitrogen oxides can be reacted with ammonia to form nitrogen and water:

$$3NO + 2NH_3 \rightarrow 5N_2 + 3H_2O$$
$$6NO_2 + 8NH_3 \rightarrow 7N_2 + 12H_2O$$

In units equipped with CO boilers, this can be accomplished by injecting ammonia into the flue gas upstream of the burners. Either anhydrous ammonia or aqueous ammonium hydroxide can be used. If the aqueous solution is used, the injection system must be designed to provide good atomization of the liquid to insure that it will evaporate in the flue gas line.

If the flue gas system does not include a fired CO boiler, then some form of catalytic reduction must be employed. Typically, this consists of passing the flue gas and ammonia through a chamber containing a catalyst on a monolithic support.

In regenerators where the spent catalyst is spread across the top of the regenerator bed (countercurrent regenerators) some of the nitrogen oxides will react with the carbon on the spent catalyst.[9]

$$2C + 2NO \rightarrow 2CO + N_2$$

Tests on commercial units have shown as much as an 80% reduction in NO_x for countercurrent regeneration.[9]

Design and Operating Considerations

FCC flue gas is in many ways similar to flue gases from coal- and oil-fired power boilers. Thus, there is a tendency to assume that flue gas clean-up systems that have proven successful in power boiler installations will work well in FCC applications. This assumption can lead to failure or poor performance of the clean-up system.

Unlike power boiler flue gas, the flue gas from an FCC regenerator is subject to large changes in composition and solids loading. Thus, flue gas scrubbers and other clean-up equipment in FCC service must be capable of operating over a wide range of conditions. The chemistry and operating parameters of new systems proposed for FCC service should be reviewed closely to insure that they will operate across all possible conditions.

New systems and systems not previously used in FCC flue gas sevice should be approached with caution. The potential operational, financial, and legal consequences of the system not performing as designed should be weighed against the possible advantages.

References

[1]Perry, R.H., Chilton, C.H., and Kirkpatrick, S.D., *Chemical Engineers' Handbook Fourth Edition,* McGraw Hill (1963), pp. 5–33.

[2]Dziewulski, T.A., and Bews, J.H., "Recover Power from FCC Units," *Hydrocarbon Processing,* December (1978), pp. 131–135.

[3]Panel Response, NPRA Cat Cracking Seminar (1996).

[4]Schiller, M., *FCCU Electrostatic Precipitator Handbook* (1990).

[5]Cunie, J.D., and Feinberg, A.S., "Innovations in FCC Wet Gas Scrubbing," NPRA Annual Meeting Paper AM-96-47 (1996).

[6]Nyman, G.B.G., and Tokerud, A., "Scrubbing by Seawater, A Simple Method of Removing SO_2 from Flue Gases," NPRA Annual Meeting (1992).

[7]NPRA Q&A Session (1992).

[8]*Hydrocarbon Processing,* November (1995), p. 145.

[9]Miller, R.B., Johnson, T.E., Santner, C.R., Avidan, A.A., and Beech, J.H., "Comparison between Single and Two Stage FCC Regenerators," NPRA Annual Meeting Paper AM-96-48 (1996).

6
Cyclones

FCC Cyclone Systems

Cyclones are used in both the reactor and the regenerator to separate catalyst from gas streams. Figure 6-1 is a schematic of a typical cyclone. Gas and solids enter the cyclone through the entrance duct. This duct is oriented tangentially to the cyclone body. Thus, the gas and solids are forced into a spiral flow path inside the cyclone.

As the solids flow along this spiral path, they move to the cyclone wall. The solids are collected by contact with the cyclone wall and these collected solids flow down the wall, into the dipleg and out of the cyclone. The gas flow reverses at the bottom of the cyclone and the gas flows up through the cyclone body and out of the cyclone through the exit duct.

Regenerator Cyclones

FCC regenerators are generally equipped with two stage cyclones (Figure 6-2). In these systems, the outlet duct of the first stage cyclone is coupled directly to the inlet duct of the second stage cyclone system. Some early regenerator designs used three stage systems. These were required due to the high attrition rates of the early FCC catalysts. The improved attrition resistance of modern catalysts have made these systems unnecessary.

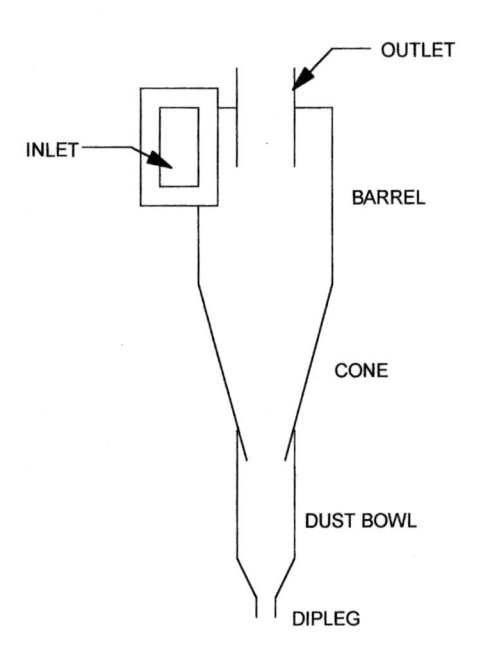

FIGURE 6-1. Typical cyclone.

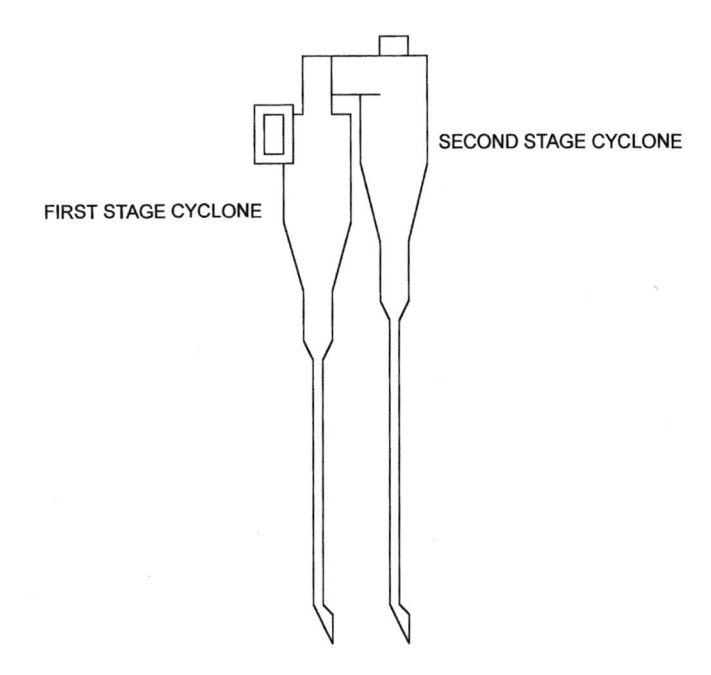

FIGURE 6-2. Two stage cyclones.

Since there is a pressure drop associated with the flow of gas through the cyclones, the pressure inside the cyclone body is less than the pressure in the regenerator. Thus, the cyclone diplegs must be sealed to prevent the flow of gas up the diplegs and into the cyclone body. If this flow were to occur, cyclone performance would be adversely affected.

Normally, the diplegs are sealed by submerging the bottom of the dipleg in the regenerator bed. When this is done, a level of catalyst is established in the dipleg and this "dipleg back up" provides a positive gas seal.

Solids load to the first stage regenerator cyclones is usually quite high, and the solids flow down the first stage dipleg is large. This flow is usually sufficient to assure a positive seal in the dipleg. Thus, the bottom of the first stage dipleg is often left open. A splash plate is suspended below the bottom of the dipleg to prevent the entrance of gas bubbles from the bed.

Since the first stage cyclones collect more than 90% of the solids entrained from the regenerator bed, the loading to the second stage is low. Thus these diplegs are lightly loaded and the flow of catalyst is often not sufficient in itself to assure a positive seal. Second stage cyclones must, therefore, be fitted with check valves at the end of the dipleg. These are generally in the form of trickle valves (Figure 6-3). These valves prevent gas from entering the dipleg and allow a catalyst level to build up in the dipleg.

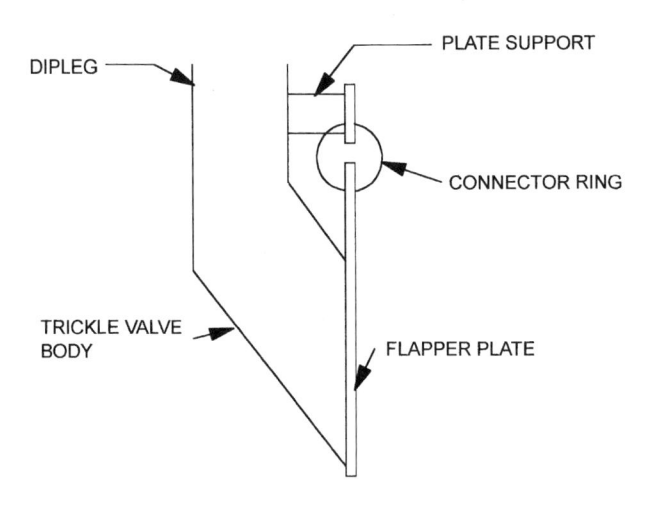

FIGURE 6-3. Trickle valve.

Reactor Cyclones

Early reactors were fitted with two stage cyclone systems similar to those used in FCC regenerators. These reactors were intended to operate as bed crackers with hydrocarbon gases flowing up through a fluidized bed. Catalyst entrainment from these fluid bed reactors was high and thus, two stage cyclones were necessary to prevent unacceptable catalyst losses to the main fractionator.

Early riser cracking units were often bed crackers in which the reactor bed level was lowered to below the riser discharge. Since these units were not equipped with any form of riser termination device, catalyst entrainment to the cyclones remained high and two stage cyclones continued to be required.

Many of the first purpose built riser crackers retained the use of two stage reactor cyclones. In many cases, this proved to be a mistake.

Even the simplest of the riser termination devices used has a separation efficiency of 50% or more. Improved inertial systems can have efficiencies of 70–80%. Thus, the catalyst load to the reactor cyclones is much lower than was the norm for older bed crackers and most of this catalyst will be removed by the first stage cyclone.

This results in a low catalyst flow to the second stage cyclone. In many cases, the amount of catalyst collected by the second stage is negligible and the second stage diplegs are seriously underloaded. When the diplegs are not sufficiently loaded it is difficult to form a positive seal and gas flows up the dipleg. This leads to erosion of the second stage cyclone and increased catalyst losses to the main fractionator.[1]

The low catalyst flow down the dipleg can also result in the accumulation of coke in the dipleg. If these coke deposits prevent the trickle valve from closing, significant gas and catalyst can flow up the dipleg.

This problem can be avoided by using a single cyclone stage in riser crackers. Often, the replacement of two stage reactor cyclone systems with a single cyclone stage actually improves collection efficiency.

Several modern designs also use cyclones as the riser termination device. These include traditional riser cyclone installations as well as the closed cyclone systems offered by various FCC licensors.

Cyclone Design

FCC cyclones are usually purchased from outside vendors. Cyclones are used for particle collection in a number of industries. As with most equipment used in FCC service, however, special design considerations are necessary to assure good performance. For this reason, cyclones should only be purchased from a vendor with a proven track record in FCC applications.

The actual cyclone designs are done using proprietary procedures. There are, however, several basic design guidelines that should be checked by the refinery. These are provided below.

Inlet Velocities, ft/sec
  Riser Cyclones   60–70
  Reactor Cyclones   60–80
  Regenerator First Stage   60–80
  Regenerator Second Stage   75–85

Outlet Velocities, ft/sec
  Riser Cyclones   70–80
  Reactor Cyclones   80–100
  Regenerator First Stage   70–80
  Regenerator Second Stage   100–200

Dipleg Mass Velocities, lb/ft^2/sec   120–130

The prospective vendors should be supplied with catalyst properties including particle density and particle size distribution, volumetric gas flow, the vessel diameter and in the case of riser cyclones, the catalyst circulation rate.

Cyclone Performance

Cyclone vendors use proprietary in-house design procedures to calculate the cyclone dust loadings and the separation efficiencies. Unfortunately, these procedures differ from vendor to vendor. Thus, vendor supplied performance data cannot be used to compare designs from different vendors. To make this comparison, the performance of various cyclone designs must be estimated using the same procedure. This begins by calculating the dust load to the cyclones.

The dust load to riser cyclones equals the catalyst circulation rate and the dust load to the reactor cyclones equals the catalyst not separated by the riser termination device (RTD).

If the RTD consists of riser cyclones, this load can be calculated as detailed below. For other devices, the dust load can be estimated by assuming a cyclone with the same efficiency as the RTD.

For regenerator cyclones or for reactor cyclones in bed crackers, the dust loading is equal to the catalyst entrainment from the bed.

Catalyst Entrainment

The catalyst entrainment to the cyclones can be estimated by the following equation[2]:

$$\log Y = 1.778 + 0.069 \log X - 0.445 (\log X)^2$$

$$Y = \frac{W}{V \rho_g}$$

$$X = \frac{V^2}{g D_p \rho_p^2}$$

where:

W = entrainment, lb/sec/ft^2
V = gas velocity at TDH, ft/sec
ρ_g = gas density, lb/ft^3
g = 32.2 ft/sec^2
D_p = particle diameter, ft
ρ_p = particle density, lb/ft^3

The entrainment calculated by the above equations is the entrainment that would be realized from a bed consisting only of particles with diameter D_p. In actuality, cracking catalyst consists of a mixture of particle sizes. Thus, the single diameter entrainment must be corrected based on the weight fraction of that particle in the mix. This is done by selecting particle size ranges and assuming that all of the particles in that range have the same diameter.

The typical ranges are:

Diameter Range, microns	Assumed Diameter, microns
0–20	10
20–40	30
40–60	50
60–80	70
80–100	90
100+	120

The fraction of catalyst particles in each range is determined from the equilibrium catalyst particle size distribution. This fraction is then multiplied by the entrainment calculated for each diameter to give the actual entrainment.

These entrainment are valid for any elevation above the transport disengaging height (TDH). The TDH is the height above the bed at which catalyst entrainment does not decrease with further increases in height. At this height localized velocity eddies caused by bubble rupture at the top of the bed have decayed and the gas velocity has stabilized.[3]

Strictly speaking, TDH is a function of the gas velocity and the bubble diameter at the top of the bed.[3] In practice, the bubble diameter is difficult to calculate. Instead, the TDH is often correlated against gas velocity and regenerator diameter. This relationship is given by:[1]

$$\log TDH_{20} = 1.312 + 0.07 (V - 3.0)$$

$$TDH = TDH_{20} + 0.1 (D - 20)$$

where:

V = gas velocity at top of bed, ft/sec
D = vessel diameter at top of bed, ft

Since the TDH is a function of the magnitude of velocity spikes created at the top of the bed, the velocity and diameter at the top of the bed are the relevant parameters for this calculation. Changes in velocity or diameter in the dilute phase have no effect on TDH.

Collection Efficiency

Cyclone collection efficiency is a function of cyclone dimensions and the cyclone inlet velocity. The relevant dimensions are:[4]

cyclone diameter	D_c
inlet height	a
inlet width	b
outlet length	S
outlet diameter	D_e
cylinder height	h
overall height	H
dust outlet diameter	B

For any particle diameter, the collection efficiency is:[4]

$$E_{Oi} = 1 - \exp\left(-2\left[\frac{G\tau_i}{D_c^3}(n+1)\right]^{0.5/(n+1)}\right)$$

In this equation:

$$\tau_i = \frac{\rho_p (D_{pi})^2}{18\mu}$$

$$G = \frac{8K_c}{K_a^2 K_b^2}$$

$$n = 1 - \left[1 - \frac{(12D_c)^{0.14}}{2.5}\right]\left[\frac{T+460}{530}\right]^{0.3}$$

where:

ρ_p = particle density, lb/ft^3
D_{pi} = particle diameter, ft
μ = gas viscosity, lb/ft-sec
T = gas temperature, °F

K_a, K_b, and K_c are determined from the catalyst dimensions.

$$K_a = \frac{a}{D_c}$$

$$K_b = \frac{b}{D_c}$$

K_c, the cyclone volume constant, is determined based on the natural length of cyclone or on the length of the cyclone below the exit duct. The natural length, l, is determined by:

$$l = 2.3 D_e \left(\frac{D_c^2}{ab} \right)^{1/3}$$

If l is less than $(H - S)$, then the volume is the cyclone volume at the natural length:

$$V = \frac{\pi D_c^2}{4}(h - S) + \left(\frac{\pi D_c^2}{4}\right)\left(\frac{l + S - h}{3}\right) \times \left(1 + \frac{d}{D_c} + \frac{d^2}{d_c^2}\right) - \frac{\pi D_e^2 l}{4}$$

If, however, l is greater than $(H - S)$, then V is the volume of the cyclone below the exit duct:

$$V = \frac{\pi D_c^2}{4}(h - S) + \left(\frac{\pi D_c^2}{4}\right)\left(\frac{H - h}{3}\right) \times \left(1 + \frac{B}{D_c} + \frac{B^2}{D_c^2}\right) - \frac{\pi D_e^2}{4}(H - S)$$

Once V has been calculated, the volume constant can be determined by:

$$K_c = \frac{2V_s + V}{2D_c^3}$$

where:

$$V_s = \frac{\pi(S - 0.5a)(D_c^2 - D_e^2)}{4}$$

Once the individual particle efficiencies have been calculated, the overall efficiency can be calculated by:

$$E_o = \sum E_{oi} x_i$$

where x_i is the weight fraction of the particles of diameter i in the dust entering the cyclone.

The calculated collection efficiencies from these equations is for a lean dust loading.[4] The actual collection efficiencies will be higher. This is due to the effects of interparticle interactions in the cyclone. The actual efficiencies for each particle can be estimated using curves found in API publication 931 Chapter 11.[5]

Operating Considerations

Pressure Drop

Koch and Licht[4] also provide an equation for cyclone pressure drop:

$$\Delta P = 0.003 \rho_g V_I^2 N_H$$

where:

ρ_g = gas density, lb/ft^3
V_I = inlet gas velocity, ft/sec

and N_H can be estimated from:

$$N_H = \frac{16ab}{D_e^2}$$

Dipleg Backup

Since there is a pressure drop through the cyclones, the inside of the cyclone normally operates at a lower pressure than the reactor or regenerator vessel (the one exception to this rule is riser cyclones, which operate at a positive pressure relative to the reactor vessel). A catalyst level must be established to prevent gas from flowing up the dipleg. This level is referred to as the dipleg backup.

The pressure at the top of the dipleg can be calculated by subtracting the pressure drop associated with the outlet duct from the overall cyclone pressure drop. The outlet duct pressure drop is due to the acceleration of the gas as it enters the duct. This can be estimated by using the following formula:[5]

$$\Delta P = 0.000108 \, KV\rho_g \, (V_e^2 - V_b^2 + KV_e^2)$$

where:

V_e = outlet velocity, ft/sec
V_b = barrel velocity, ft/sec
ρ_g = gas density, lb/ft^3

The factor K is based on the area ratio between the cyclone barrel and the cyclone outlet using the following:

Area Ratio	K
0	0.50
0.1	0.47
0.2	0.43
0.3	0.395
0.4	0.35

The pressure at the bottom of the dipleg is equal to the reactor pressure plus the static head of the catalyst bed if the dipleg is submerged. The dipleg back up is the level of catalyst required to balance these pressures. For the purpose of this calculation, the density of the catalyst in the dipleg is assumed to be 25 lb/ft^3. This low pseudo density is used due to the fact that the catalyst in the dipleg is generally not well fluidized.

The actual catalyst level in the dipleg will rise and fall to adjust for transient pressure changes in the pressure differential between the cyclone and the vessel. If the dipleg is sealed in a fluid bed, the level will also adjust automatically for changes in bed density.

To allow for these changes, the actual length of the dipleg above the bed should be longer than the calculated dipleg backup.

Erosion

Erosion is a common problem in cyclones. High velocities and high particle loadings subject the cyclone bodies and internals to considerable wear. Over time, this severe environment often causes damage.

The most common locations for erosion are in the secondary cyclone, especially in the diplegs, and the lower portion of the cone.[1] While some erosion is normal, severe erosion and frequent failures at any one point generally indicate problems with design or operation.

Serious erosion on the inlet ducts, crossover ducts or outlet ducts generally indicates excessive velocities at these points. This is often caused by operational changes that increase the gas flow (increased feed rates, increased coke burn, lower pressures). If the high velocities cannot be corrected by operational modifications, then it may be necessary to modify or replace the cyclones.

Erosion problems in the dipleg or at the bottom of the cyclone are generally an indication of gas flow up the dipleg.[1] This may indicate an oversized dipleg, or it may be due to problems with the dipleg trickle valves. Erosion in the bottom of the cyclone or at the top of the dipleg may also indicate gas flow up the dipleg. It may also, however, be due to impingement of the cyclone vortex on the bottom of the cyclone. This is generally due to increased cyclone velocities (either outlet or inlet) and may be due to operational changes or in some cases to modifications made to increase the efficiency of the cyclone.

Repair vs. Replacement

The decision as to whether to repair or replace damaged cyclones is generally based on costs. As long as the cyclone body is sound, damaged cyclones can be repaired and returned to service. The time required to make these repairs is often significant, however, especially if the cyclone lining must be replaced. In these cases it may be more cost effective to replace the cyclones.

This is especially true if the cyclones have experienced repeated failures at the same point. This indicates that the existing cyclone design is inadequate for the current service. In this case, replacement with an improved design will often result in significant long-term savings.

References

[1]Tenney, E.D., "FCC Cyclone Problems and How They Can Be Overcome with Current Designs," Grace–Davison FCC Technology Conference, June (1992).

[2]Ewell, R.B. and Gadmer, G., "Design Cat Crackers by Computer," *Hydrocarbon Processing,* April (1978), pp. 125–134.

[3]Zenz, F.A., "Particulate Solids: The Third Fluid Phase in Chemical Engineering," *Chemical Engineering,* September 28 (1983), pp. 60–67.

[4]Koch, W.H. and Licht, W., "New Design Approach Boosts Cyclone Efficiency," *Chemical Engineering,* November 7 (1977), pp. 80–88.

[5]API Publication 931 Chapter 11 (1975), pp. 11–32.

7
Fluidization and Standpipe Flow

As the name implies, fluid catalytic cracking is based on maintaining the solid catalyst in a fluid state. Fluidization and fluid particle systems are complex and difficult to understand. In the fifty-plus years since the introduction of catalytic cracking, there has been considerable research on fluidization in both academia and industry. These efforts have lead to increasingly complex models designed to address the many variables involved in fluid particle systems.[1]

Most of these discoveries go beyond the practical needs of engineers involved in the design or operation of fluid catalytic cracking units. A basic understanding of fluidization principles is necessary, however, to have a full understanding of the operation of the FCC. A somewhat more detailed understanding is often required to diagnose and correct commonly encountered catalyst circulation problems.

Fluidization Fundamentals

The behavior of a fluid bed will depend on the nature of the solids in the bed and on superficial gas velocity in the bed. The bed will generally go through several distinctive fluidization states as the gas velocity increases. The nature of these states and the velocities at which they occur are an important parameter in understanding fluidization performance of the solids.

Fluidization States

When a gas stream is passed upward through a bed of solid particles, the bed goes through several distinct states. The nature of these fluidization states and the velocities at which they occur depend on the properties of the solid particles and, to a lesser extent the properties of the gas.

Geldart[2] developed a particle classification scheme that grouped particles into four groups—A,B,C and D. Each of these groups shows specific behavior as the gas flow is increased above that required for minimum fluidization.[3]

Group A—Group A particles are referred to as aeratable particles. FCC catalyst is a Group A solid. Beds made up of Group A solids continue to expand without the formation of bubbles as gas velocity is increased above minimum fluidization.[3] This expansion continues over a certain range of velocities until the minimum bubbling velocity is reached. At this point, bubbles begin to form. At minimum bubbling, the bed actually contracts slightly as some of the gas moves into the bubbles.

Group B—Group B particles are called granular particles. They are larger and denser than Group A particles. Sand is a typical Group B solid. Beds made up of Group B particles begin to bubble as soon as they are fluidized.[3]

Group C—Group C contains the smallest and lightest particles classified by Geldart. Group C particles are referred to as cohesive. Beds containing these particles are dominated by interparticle forces.[3] FCC catalyst fines are an example of a Group C solid. Because these particles are cohesive, they are difficult to fluidize and gas channeling through the bed is common.

Group D—Group D, generally referred to as spoutable particles, consists of large dense particles. Group D particles are difficult to fluidize and generally show high minimum fluidization velocities. Additionally, particle beds made up of Group D particles often spout at or near minimum fluidization velocity.[3]

As discussed above, FCC catalyst is a Group A or aeratable powder, and it is the fluidization characteristics of the group that are of importance to FCC engineers.

Group A particles generally go through five fluidization stages as the gas velocity is increased. Initially, the bed remains unfluidized or packed. In this state, the gas flows through the spaces between the catalyst particles. Since the bed is fixed, the area available for flow is fixed. Thus, the pressure drop through the bed increases with the square of the gas velocity.

When the pressure drop through the bed equals the weight of the bed per unit area, the bed lifts and becomes fluidized. The velocity at which this occurs is referred to as the minimum fluidization velocity or U_{mf}. As the gas velocity is increased above U_{mf} the bed continues to expand. This continues with increasing velocity until the minimum bubbling velocity, U_{mb}, is reached.[3]

At U_{mb} bubbles begin to form in the bed. As the bubbles form, they actually draw gas from the fluidized bed. Thus, the bed actually contracts slightly at U_{mb}. Above the minimum bubbling velocity, additional gas added to the bed travels upward in the form of bubbles. As the gas velocity is increased, more bubbles are formed and the total bed volume increases.

As gas velocities are increased further, the bubbling bed gives way to a turbulent bed. In a turbulent bed, bubbles are no longer present. Instead the gas flows upward through a bed consisting of densely packed catalyst clusters. These clusters move about rapidly and freely, and this results in the appearance of a homogeneous and highly turbulent bed.[4]

Further increases in velocity increase the turbulence in the bed as well as the entrainment of catalyst from the bed until it is no longer possible to identify separate dense and dilute phases. At this point, the bed has entered the fast fluid bed state.

Fast fluid beds differ from turbulent beds in two ways. First, as mentioned above, in a fast fluid bed, there is no clearly defined interface between a dense bed and a dilute phase. Secondly, in a turbulent bed, the bed density is largely a function of gas velocity and is essentially independent of the rate at which catalyst is added to or removed from the bed. In a fast fluid bed, however, catalyst hold-up and thus bed density is a function of both gas velocity and catalyst mass rate.[4]

The fast fluidized bed state exists until the gas velocity reaches the choke velocity. As discussed in Chapter 3, the choke velocity is a function of the particle properties, the gas properties, the particle mass velocity and the bed diameter. Once the choke velocity is reached, the system goes from a dense phase, fast fluid bed to dilute phase. The onset of this transition is sudden and is characterized by a rapid decrease in density as the solids become dispersed in the upward flowing gas.

Fluidization Curves

The fluidization properties of FCC catalyst are generally shown in the form of a laboratory fluidization curve. This curve is produced using a fluidization tube such as the one shown in Figure 7-1. A known mass of catalyst is placed in the tube and the flow of air to the bottom of this bed is gradually increased. The pressure drop across the bed and the bed height are measured as the air flow is increased. The true bed density is calculated from the measured height of the bed and the known cross sectional area of the tube. A pseudo density is also calculated by dividing the pressure drop across the bed by the bed height. Both of these densities are plotted against the log of the superficial gas velocity.

A typical set of fluidization curves is shown in Figure 7-2. Initially, the gas velocity is below minimum fluidization velocity and the true bed density does not change. The pseudo density, however, increases as the gas velocity increases. The pseudo density equals the true bed density at the minimum fluidization velocity.

Once the bed becomes fluidized, both the true bed density and the pseudo density decrease with increasing velocity as the bed expands. The pseudo density often, however, is lower than the

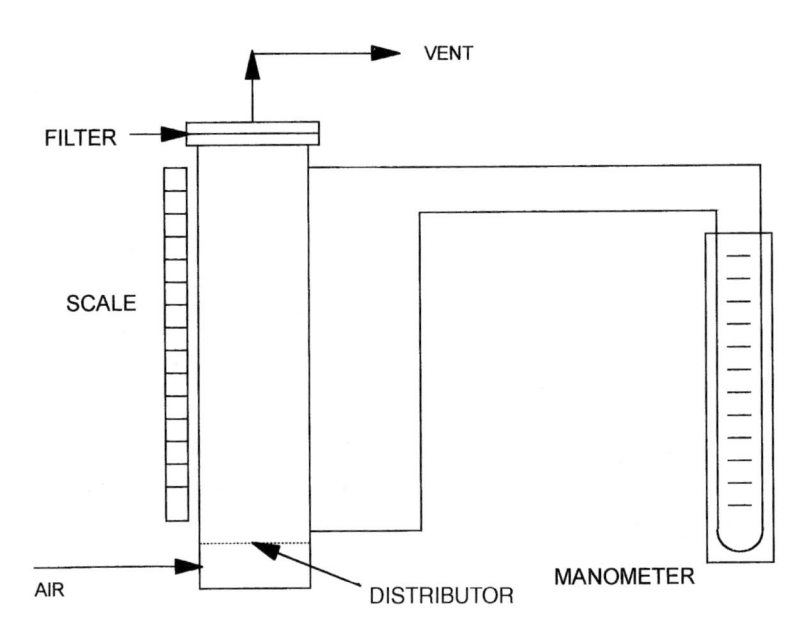

FIGURE 7-1. Fluidization tube.

true density. This may be due to a stratification of the various particle sizes in the bed, or may be due to imperfect distribution of the fluidizing gas. No bubbles are present in this homogeneous fluidization state.

The bed level can be seen to fall at the minimum bubbling velocity, and thus, the true density actually increases at the onset of bubbling. Generally, the pseudo density moves toward and coincides with the true density soon after the onset of bubbling.

As the gas velocity is increased above minimum bubbling, the bed density decreases as volume of the air bubbles increases. As the bubbling intensity increases, the pseudo density shows increasing short-term fluctuations due to the variability in the amount of gas flowing upward in the bubbles.

At the onset of turbulent fluidization, these fluctuations in pseudo density decrease significantly as the bubbles disappear.[4] In the turbulent regime, the bed density falls more rapidly with increasing gas velocity than in the bubbling bed regime.

The simple fluidization tube shown in Figure 7-2 cannot be used for measurements in fluidization states beyond turbulent fluidization. For these higher velocity regimes a circulating test unit

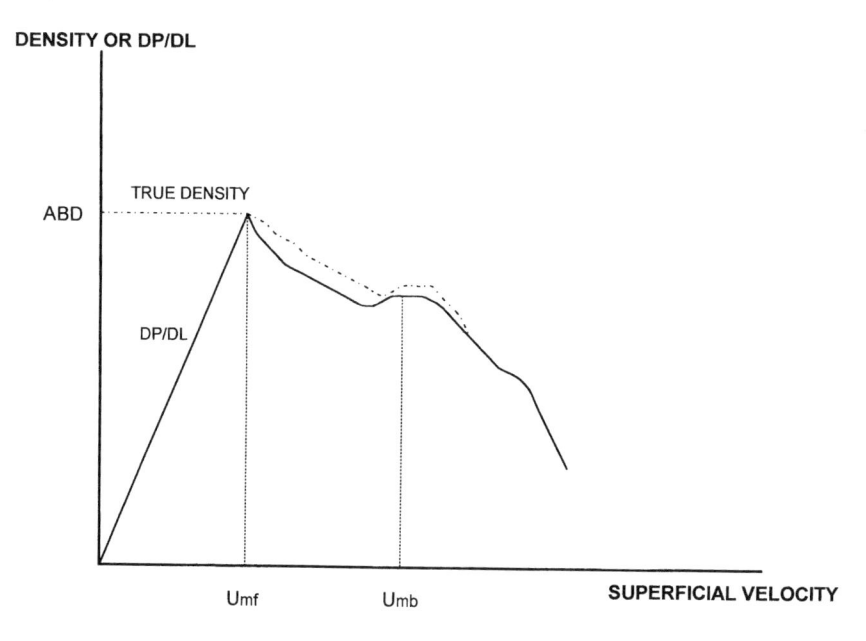

FIGURE 7-2. Fluidization curve.

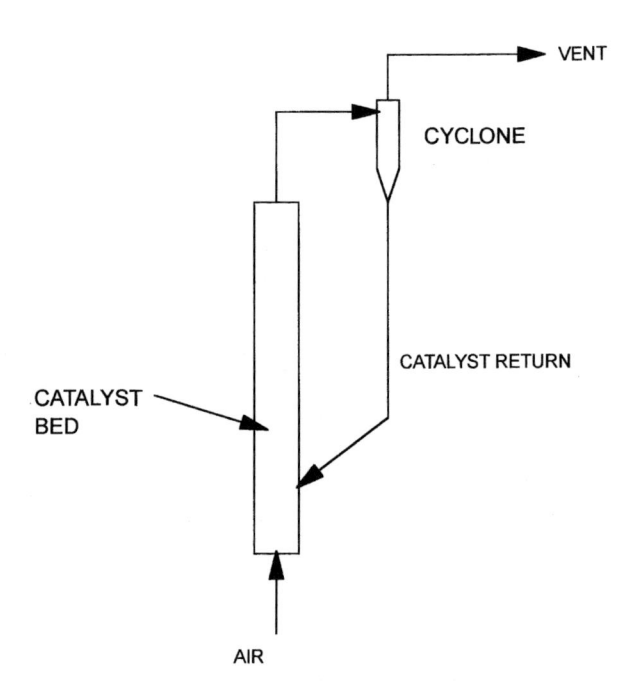

FIGURE 7-3. Circulating test unit.

such as shown in Figure 7-3 must be used. In these units, the catalyst circulates from the bottom of the test bed, up to the cyclones where it is collected and returned to the bottom of the bed. Data from these circulating test units is usually plotted using the pseudo bed density and the slip velocity.[4] The slip velocity is defined as the difference between the gas velocity, V_g and the catalyst velocity, V_p. A typical plot is shown in Figure 7-4.

The multiple curves based on different catalyst mass velocities are due to the effects of wall friction and particle acceleration.[5] If true density were used instead, a single curve would be obtained.

Fluidization in FCC Operations

Catalyst Fluidization Properties

The minimum fluidization velocity for FCC catalyst is normally in the range of 0.01–0.02 ft/sec. Minimum bubbling veloc-

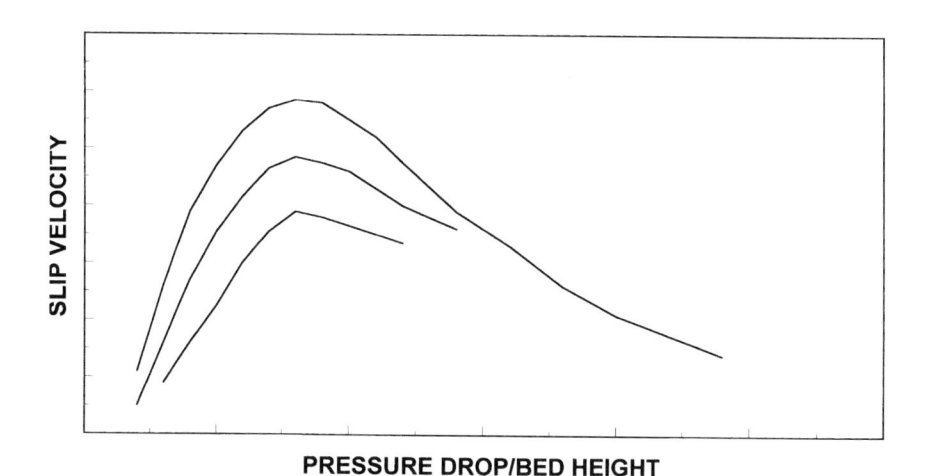

FIGURE 7-4. High velocity fluidization curves.

ities are normally 2.5–3.5 times higher than U_{mf}. The transition from bubbling to turbulent beds normally occurs at about 2.0 ft/sec. Depending on the catalyst mass velocities, a fast fluid bed can be initiated anywhere between 5.0 and 7.0 ft/sec superficial velocity. Dilute phase transport normally occurs above 10 ft/sec.

From these typical transition points, we can see that most of the catalyst fluidization states can exist in an operating FCC. Spent catalyst strippers normally operate in the bubbling bed regime. Conventional regenerators are generally operated at superficial velocities of 2.0–4.0 ft/sec and are thus turbulent beds. UOP high efficiency regenerators are designed to operate as fast fluid beds. Fast fluid beds also generally exist in the reaction riser below elevated feed nozzles and in short catalyst lift lines such as those found in Model IV FCCs. Ideally, catalyst standpipes operate in the homogeneous fluidization state between the minimum fluidization and minimum bubbling velocities.

The fluidization and minimum bubbling velocities for FCC catalyst are important parameters in evaluating catalyst fluidization. These properties are best determined by fluidization tests as described above. These tests are not, however, normally run in refinery laboratories, so it is often necessary to estimate these two critical parameters based on catalyst and gas properties.

Baeyens and Geldart[6] proposed the following equation for U_{mf}

$$U_{mf} = \frac{9 \times 10^{-4}(\rho_p - \rho_g)^{0.934} g^{0.934} d_p^{-1.8}}{\mu_g^{0.87} \rho_g^{0.066}}$$

and Abrahamson and Geldart[7] proposed the following for U_{mb}:

$$U_{mb} = \frac{2.07 e^{0.716 F} d_p \rho_g^{0.06}}{\mu_g^{0.347}}$$

where:

d_p = average particle size, m
F = wt. fraction of particles less than 45
g = gravitational constant, 9.81 m/sec^2
U_{mb} = minimum bubbling velocity, m/sec
U_{mf} = minimum fluidization velocity, m/sec
ρ_g = gas density, kg/m^3
ρ_p = particle density, kg/m^3
μ_g = gas viscosity, kg/m-sec

Densities and Pressure Drops

As discussed above, the density of a catalyst bed can be calculated in two ways. The true density is based on the mass of catalyst and vapors in the bed and the actual bed volume. This can be expressed as:

$$\rho_{bed} = (1 - \varepsilon)\rho_p + \varepsilon \rho_g$$

where:

ρ_{bed} = true bed density
ε = bed void fraction
ρ_g = gas density
μ_p = particle density

Since the catalyst particle density is much greater than the gas density at FCC operating conditions, this equation can be simplified to:

$$\rho_{bed} = (1 - \varepsilon)\rho_p$$

The pseudo bed density is calculated based on a measured pressure drop across a known length of the bed:

$$\rho_{pb} = \frac{\Delta P}{\Delta L}$$

where:

ρ_{pb} = pseudo bed density
ΔP = pressure difference over known bed height
ΔL = known bed height

The pseudo bed density is in actuality a measure of the pressure drop in the gas stream flowing up through the bed. In many cases, this value will be different than the true bed density. This is especially true in catalyst standpipes where the difference between the true density and the pseudo density can lead to operational difficulties.

For most FCC applications, the pseudo density is of the most interest. This "density" actually determines the pressures found in fluidized beds and thus, the unit pressure balance. For fluid beds, this density can be estimated from the superficial bed velocity and the catalyst particle density using Figure 7-5. These curves are valid between U_{mb} and the transition to a fast fluid bed.

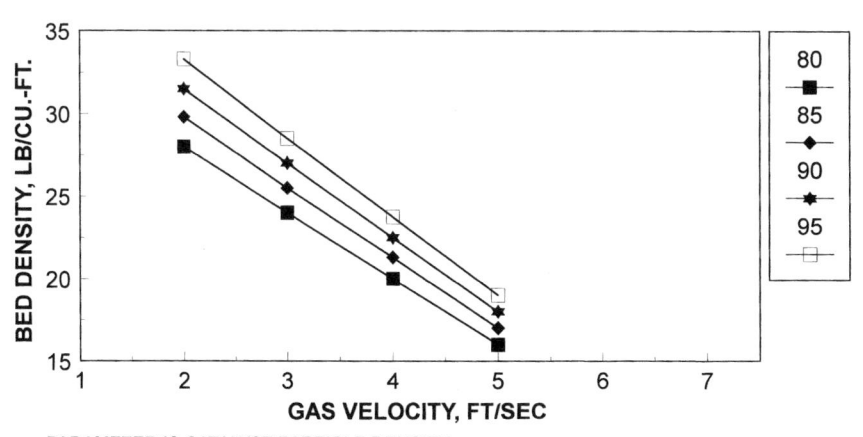

FIGURE 7-5. Fluidized bed densities.

Figure 7-5 can also be used to calculate the true density in a fast fluid bed. The pseudo density will, however, be higher since it will also include the pressure drop from catalyst acceleration. This additional pressure drop can be estimated from:

$$\Delta P_a = \frac{W \Delta V_p}{g_c}$$

where:

ΔP_a = pressure drop due to catalyst acceleration, lb/ft^2
ΔV_p = change in catalyst velocity, ft/sec
g_c = 32.2
W = catalyst mass velocity, lb/ft^2-sec

This is the same as the catalyst acceleration term used in the equation for riser pressure drop presented in Chapter 3. The catalyst velocity, V_p, in the fast fluid bed is calculated using the gas velocity and the choke velocity as discussed in that chapter. The pseudo density of the fast fluid bed can now be estimated by:

$$\rho_{pb} = \rho_{bed} \frac{\Delta P_a}{\Delta L}$$

where:

ΔL is the height of the fast fluid bed.

Catalyst Densities

As the equations above show, catalyst properties, especially catalyst particle densities play a major role in estimating the properties of fluid beds. The catalyst particle density can be calculated from the following equation:

$$\rho_p = \frac{\rho_{sk}}{(\rho_{sk} \times PV) + 1}$$

where:

ρ_p = catalyst particle density, g/cc

ρ_{sk} = catalyst skeletal density, g/cc
PV = catalyst pore volume, cc/g

The catalyst skeletal density is the mineral density of the solid portion of the catalyst and does not include the volume of the catalyst pores (this volume is included in the particle density). The skeletal density is determined by the composition of the catalyst particle and can be estimated based on the catalyst alumina content:

$$\rho_{sk} = \frac{1}{\frac{Al}{3.4} + \frac{Si}{2.1}}$$

where:

Al = weight fraction alumina in catalyst
Si = weight fraction silica in catalyst $(1.0 - Al)$

Standpipes

Catalyst standpipes are a key ingredient in fluid catalytic cracking units. They are also probably the least well understood. While fluidization problems can occur in other parts of the FCC, these can usually be traced to either poorly designed or damaged equipment. Standpipe flow problems on the other hand are often the result of operational difficulties.

A firm understanding of standpipe flow principles is necessary to deal with these problems. Unfortunately, there is considerable folklore concerning standpipes, some of which is misleading.

Standpipes are frequently thought of as being filled with a liquid. This is an oversimplification. In fact, standpipes are complex systems involving interactions between the solid catalyst, fluidizing gases, and the standpipe itself.

Standpipe Principles

Standpipes are used to transport catalyst against a pressure gradient. That is, the standpipe moves the catalyst from a low pressure region to one of higher pressure. This can only be achieved if there is a flow of gas in the standpipe. This gas flow must be upward relative to the flow of catalyst.[8] The direction of gas flow relative to the standpipe walls may, however, be either upward or downward.

The velocity of the gas relative to the catalyst is the slip velocity, V_s, and is defined as:

$$V_x = V_g - V_p$$

where:

V_s = slip velocity
V_p = catalyst velocity
V_g = gas velocity

By convention, the direction of V_p is normally taken as positive. As long as V_g is less than V_p, the slip velocity will be negative; that is, the relative gas flow will be in the direction opposite to the flow of solids.

Gas flowing in the opposite direction relative to the catalyst produces a frictional pressure drop in the direction of relative gas flow—the direction opposite to the flow of catalyst.[8] When viewed in the direction of catalyst flow this frictional pressure drop is seen as a pressure increase. The pressure difference per unit length of the standpipe, $\Delta P/\Delta L$, is the same as that in the pseudo density fluidization curve (Figure 7-2).

This illustrates two of the most common misconceptions concerning standpipe operation. The first of these is the belief that the standpipe must be fluidized to produce a pressure buildup. As Figure 7-2 shows, gas flowing upward through a packed bed will produce a pressure gradient across the bed and thus, a positive pseudo density.

It is possible to operate a standpipe as a packed bed. In these standpipes, the slip velocity is less than the minimum fluidization velocity. Packed bed flow is not used in FCC catalyst standpipes since the catalyst mass velocities are low. Packed bed flow does exist, however, in some cyclone diplegs.

The second major misconception relates to the calculation of standpipe density. Since $\Delta P/\Delta L$ has the same units as density, this measure is often treated as the actual density in the standpipe. In fact, this would only be true if the standpipe were either exactly at minimum fluidization or vigorously bubbling. In truth, most standpipes operate with slip velocities between minimum fluidization and minimum bubbling or just slightly above minimum bubbling. Under these conditions, the pseudo density may not be equal to the actual density.

In addition, in an operating standpipe, the flow of catalyst and gas down the standpipe creates a frictional pressure loss in the direction of catalyst flow. This pressure loss reduces the observed pressure buildup and thus, reduces the pseudo density calculated from the pressure change over the length of the standpipe.

Cold flow studies by Wrench et al.[9] have indeed shown that "The actual standpipe density is close to the bulk density of the catalyst. The density calculated from the standpipe pressure buildup will be less than this value." This misconception with regard to standpipe densities has resulted in many problems in standpipe design and operation.

Design and Operation

Given the misconceptions and folklore that surround standpipe design and operation, it is not surprising that they often appear more art than science. This can be especially frustrating to refinery engineers when faced with standpipe problems. The basic principles of design and operation are, however, straight forward and easy to understand.

Design. The vast majority of standpipes in FCC service are designed as underflow standpipes (Figure 7-6). An underflow standpipe catalyst draws catalyst from the interior of a fluidized bed. The standpipe is operated completely full of catalyst and the flow of catalyst is controlled by a valve (usually a slide valve) located at the bottom of the standpipe.

Model IV and Exxon Flexicracking units utilize an overflow standpipe (Figure 7-7). In an overflow standpipe, the catalyst is drawn from the top of the fluidized bed. This catalyst overflows from the bed into the standpipe. The standpipe itself is only partially filled. In operation, the level in the standpipe changes to adjust the pressure buildup in the standpipe. There is no catalyst control valve at the end of an overflow standpipe; instead, the catalyst flow is regulated by adjusting the pressure difference

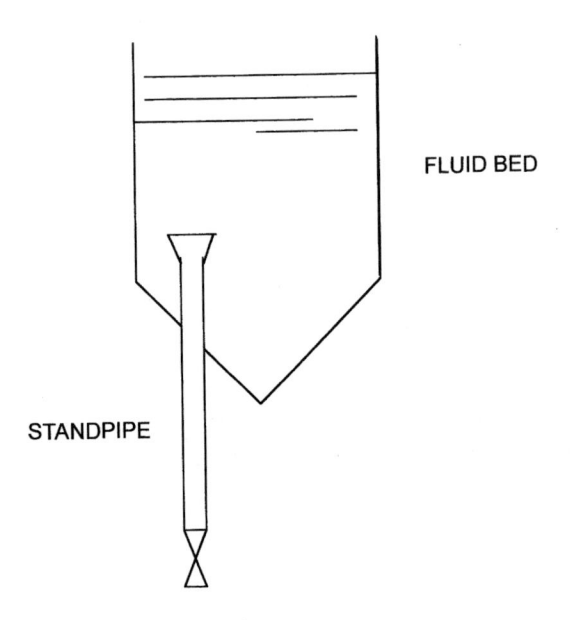

FIGURE 7-6. Underflow standpipe.

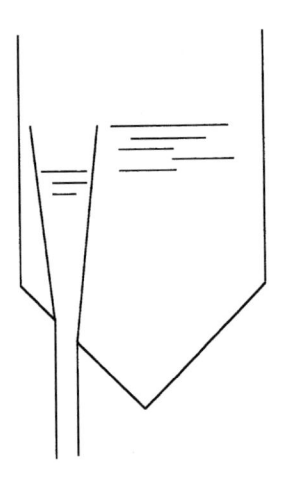

FIGURE 7-7. Overflow standpipe.

between the top and bottom of the standpipe. In addition to the catalyst standpipes in Model IV and Flexicracking units, cyclone diplegs also operate as overflow standpipes.

Regardless of the type of standpipe used, the basic principles of design and operation are the same. Standpipe size is set by the

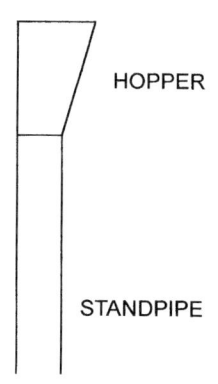

FIGURE 7-8. Conical hopper.

catalyst mass velocity. Typical design values are 180–200 lb/ft²-sec. Acceptable operating values vary between 135–220 lb/ft²-sec. Operation outside of this range, while possible, is usually difficult and subject to flow instabilities.

Standpipes that draw catalyst from an active bed are normally equipped with conical hoppers (Figure 7-8). The top of the hopper is sized with a diameter equal to 2.0–2.5 times the standpipe diameter and the angle of the hopper cone is between 10 and 30 degrees off of the vertical. The purpose of these hoppers is to deaerate the catalyst from bed density to standpipe "density" and thus prevent the entrainment of bubbles into the standpipe.

Model IV and Flexicracker overflow standpipes are equipped with a similar hopper often refered to as the overflow well. Early Model IV designs used a cylindrical overflow well followed by a conical transition to the standpipe diameter. Later designs have used a long tapered overflow well.

Proper design of these hoppers is critical to avoid operational problems in the standpipe. A poorly designed hopper can result in either the entrainment of bubbles into the standpipe or defluidization of the catalyst in the hopper. Either of these will cause the flow of catalyst in the standpipe to become unstable.

Standpipes that draw catalyst from the active bed can also be affected by instabilities in the bed. These instabilities can be transmitted to the hopper and standpipe where they will cause density fluctuations in the catalyst entering the standpipe.

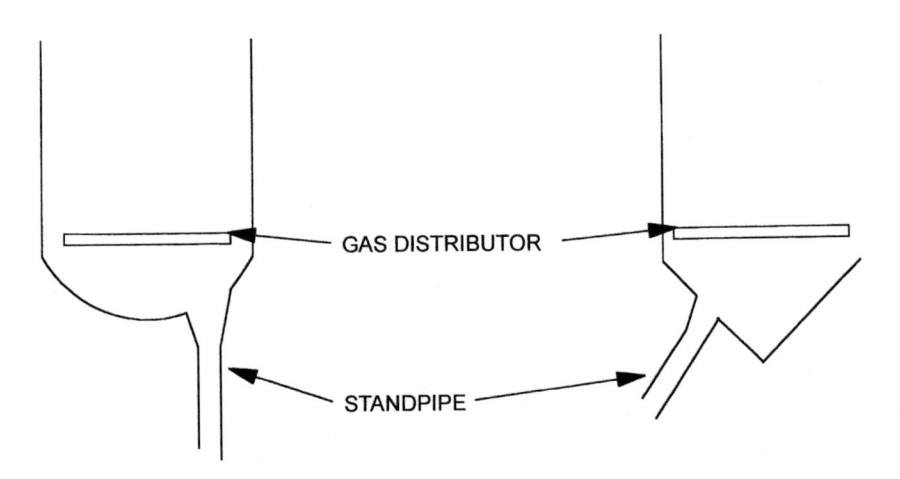

FIGURE 7-9. Standpipe entry below gas distributor.

Standpipes drawing from below the main gas distributor (Figure 7-9) do not suffer from these problems and the design of the hopper is less critical.

Direction changes in standpipes should be avoided wherever possible. As the catalyst/gas mixture flows trough a directional change, there is a tendency for the gas to separate from the catalyst. This can lead to a large stationary gas bubble over a region of defluidized catalyst. This will act as a flow restriction in the standpipe and reduce both the flow of catalyst as well as the standpipe pressure buildup.

Operation. The catalyst flowing into the standpipe entrains vapor from the catalyst bed. If the standpipe entrance is properly designed, this entrained vapor will be sufficient to produce a slip velocity close to the minimum bubbling velocity. Excessive vapor entrainment can lead to bubble formation at the standpipe entrance. These bubbles can restrict the flow of catalyst into the standpipe. If the catalyst entering the standpipe is drawn from a poorly fluidized bed, insufficient vapor may be entrained into the standpipe and this can lead to low catalyst mass velocities.

As the catalyst/gas mixture flows down the standpipe, the increase in pressure compresses the gas in the spaces between the catalyst particles as well as the gas in the catalyst pores. This results in a decrease in the gas volume and thus a decrease in the relative velocity between the gas and the catalyst.

If the standpipe is long enough, this compression will continue until the slip velocity falls below U_{mf} for the catalyst. When this occurs, the standpipe defluidizes. This results in a significant decrease in catalyst circulation.

The tolerance of a particular catalyst to compression is based on the ratio of the U_{mb} to U_{mf}. This can be calculated from the previously presented equations for these parameters:

$$\frac{U_{mb}}{U_{mf}} = \frac{2300 \rho_g^{0.126} \mu^{0.523} e^{0.716F}}{d_p^{0.8} g^{0.934} (\rho_p - \rho_g)^{0.934}}$$

Wrench et al.[9] reported that this ratio—referred to as the G factor—provided an excellent measure of the catalyst's flow performance. Raterman[10] published a similar parameter based on the ratio of the deaeration velocity to the minimum fluidization velocity.

The ratio of density at minimum bubbling to the density at minimum fluidization can be calculated based on the G factor:[11]

$$\frac{\rho_{mb}}{\rho_{mf}} = G^{-0.22}$$

Since the density at minimum fluidization is equal to the ABD of the catalyst, the density at minimum bubbling is given by:

$$\rho_{mb} = (ABD)G^{-0.22}$$

Thus, a high G factor indicates a larger difference between the density at the standpipe entrance and the density at which the catalyst in the standpipe will defluidize. Higher G values indicate a catalyst with greater tolerance for compression and these catalysts are likely to give smoother standpipe operation.

Since the pressure gain in long standpipes is usually sufficient to compress the gas to the point of catalyst defluidization, these standpipes are generally equipped with supplemental aeration. The purpose of these aeration systems is to supply just enough gas volume to the flowing mixture to return it to the standpipe entrance density.

The volume of aeration required at each point is a function of the standpipe density at the standpipe entrance, the distance between the entrance and the first aeration point (or, for subsequent aeration points, the distance between aeration points) and the pressure buildup in the standpipe.

The volume of gas per cubic foot of standpipe volume is equal to the void fraction. Therefore, the volume of gas per pound of catalyst is equal to (ignoring the mass of the gas in the standpipe):

$$\frac{ft^3_{gas}}{lb_c} = \frac{\varepsilon_{sp}}{\rho_{sp}}$$

where:

ε_{sp} = standpipe void fraction
ρ_{sp} = standpipe density

but the standpipe density is also a function of the void fraction:

$$\rho_{sp} = \rho_{sk}(1 - \varepsilon_{sp})$$

or,

$$\varepsilon_{sp} = \frac{\rho_{sk} - \rho_{sp}}{\rho_{sk}}$$

thus,

$$\frac{ft^3_{gas}}{lb_c} = \frac{\rho_{sk} - \rho_{sp}}{\rho_{sk}\rho_{sp}}$$

The catalyst skeletal density is used instead of the particle density since the gas in the catalyst pores is also subject to compression.

Obviously, the proper standpipe density must be used to calculate the volume of gas in the standpipe. The common assumption that the pseudo density calculated from the standpipe pressure increase equals the true standpipe density will almost always lead to an overestimation of the gas volume in the standpipe.

The proper density to use in this calculation is the density at the minimum bubbling velocity. This can be estimated based on the G factor as shown above or can be taken from the true density fluidization curve for the catalyst.

The volume of gas per pound of catalyst calculated above multiplied by the catalyst circulation rate will give the gas volumetric flow rate at the top of the standpipe. Shrinkage in this gas volume as the catalyst flows down the standpipe can be calculated from the change in pressure. Specifically, the loss in gas volume at the first aeration point will be:

$$\Delta ft_{i-1}^3 = ft_i^3 \left(1 - \frac{P_i}{P_1}\right)$$

where the subscripts i and 1 indicate conditions at the standpipe entrance and at the first aeration point respectively. In making this calculation, the pressures should be based on the expected pressure buildup in the standpipe; this can be estimated using a pseudo density of 33–37 lb/ft.

The gas volume lost to compression is the volume that must be replaced by the aeration gas injected at this point. Similarly, the required aeration gas rate for each subsequent aeration point can be estimated using the same formula and substituting the appropriate pressure conditions.

The use of the correct standpipe densities in this calculation eliminates the need for the often used 70% of theoretical aeration rule.

This rule is based on information published by Wrench et al. in 1985.[9] This study found that pressure buildup in a cold flow standpipe model increased with increasing aeration air up to approximately 70–75% of the theoretical aeration rate (based on an assumed standpipe density of 37 lb/ft^3. Above this aeration rate, however, the standpipe became overaerated and catalyst flow became unstable.

At overaerated conditions, large voids formed in the standpipe. Catalyst flow through these voids was in free fall and there was little or no pressure buildup. In extreme cases of overaeration, the entire standpipe was in free fall.

The test standpipe was 35 feet long with a 4 inch diameter. Individual aeration points were located every three feet along the standpipe length. Catalyst flow was regulated by a valve at the base of the standpipe.[9]

During the tests, aeration air flow was measured using rotometers and the catalyst flow was measured based on the time required to fill a measuring chamber of known volume. Standpipe pressures could be taken at each aeration point.

As quoted earlier, this actual density measured in the standpipe was much higher than either the assumed density or the measured pseudo density, and was close to the catalyst ABD. Higher densities would result in a higher percent of theoretical. Using the calculation procedure above, the percent theoretical aeration for several densities can be calculated (skeletal density is assumed to be 155 lb/ft^3):

Density, lb/ft^3	% Theoretical
36	70
40	80
45	95
50	110

Thus, if the true density in the standpipe were between 45 and 50 lb/ft^3 the aeration rate would be close to 100% theoretical. For most catalysts, these are reasonable densities for operation between minimum bubbling and minimum fluidization. Thus, the often quoted 70% rule is in fact a correction required due to the use of the wrong standpipe density in the aeration calculation. When the correct density is used, the calculated aeration requirement will be close to the optimum requirement.

Wrench et al.[9] also reported the following observations:

1. The highest pressure buildup in the standpipe resulted from evenly distributed aeration.
2. Standpipes sloped up to 15 degrees from vertical showed the same performance as vertical standpipes.
3. Standpipe flow is affected by catalyst properties.

Standpipe Problems

The following symptoms usually indicate a problem in standpipe operation.[11]

1. Low standpipe pressure buildup (pseudo density less than 33 lb/ft^3)

2. Erratic slide valve pressure differentials
3. Low catalyst circulation rates at large slide valve openings or inability to control catalyst circulation with changes in slide valve opening
4. Sudden interruptions in catalyst circulation (usually indicated by sudden drops in riser outlet temperature)
5. Large physical movements of the standpipe and/or standpipe expansion joints

These problems are generally quite disturbing when they occur and often threaten the continued operation of the FCC. Due to the seriousness of these problems, the first approach at troubleshooting often involves significant changes to standpipe operations without the benefit of proper analysis.

This can often lead to actions that aggravate rather than improve the situation. This is especially true with respect to changes in standpipe aeration rates.

The physical symptoms of overaeration and underaeration are the same: low pressure buildup and low slide valve pressure differential. This similarity masks totally different problems.[9]

If the standpipe is underaerated, the slip velocity is low, and thus the pressure buildup will low. In addition, frictional losses in the standpipe will be high and this will further decrease the pressure buildup.

If, on the other hand, the standpipe is overaerated, the pressure buildup will be low due to the presence of bubbles. These bubbles restrict the flow of catalyst.

The usual approach to low slide valve differentials is to increase the aeration rate. If the standpipe is underaerated, this may correct the problem. If, however, the standpipe is not underaerated or if the underaeration exists only in a short section of the standpipe, a general increase in the aeration rate can actually result in severe overaeration.

At this extreme, large bubbles can completely fill the standpipe. Slide valve pressure drop will decrease severely, often to a dangerously low level. Once this occurs, recovery can be difficult. Reduction in aeration rates to the previous levels will often not be sufficient. More drastic action including diverting feed and closing the slide valve may be necessary before the standpipe can be refilled.[9]

To avoid problems caused by inappropriate corrective actions, standpipe problems should be analyzed before making any operational changes.

The first step is to calculate the required aeration rate using the procedure discussed above. The required aeration rate should be calculated for each aeration point and this value should be compared to the actual volume of aeration gas injected. If the actual rate at each point differs significantly from the theoretical rate, it should be adjusted. In many cases, however, the individual aeration flows cannot be controlled independently. In these cases, the total aeration gas flow should be adjusted to minimize the possibility of overaeration.

If proper aeration does not improve standpipe operation, the next step is to measure the standpipe pressure profile. This should be done using a single, calibrated pressure gauge. Pressures should be taken at each aeration point.

These data should be plotted to show the pressure in the standpipe as a function of the standpipe elevation. Figure 7-10 shows a typical pressure profile for a well-operated standpipe. Significant variations from this profile would indicate problem areas.

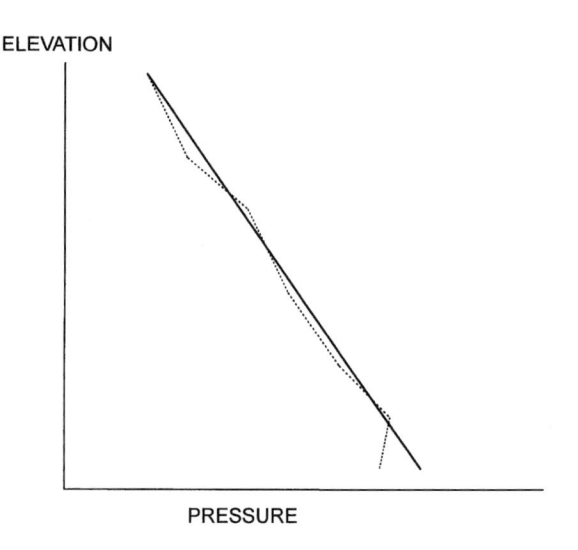

SOLID LINE INDICATES IDEAL PROFILE

FIGURE 7-10. Good standpipe pressure profile.

Once these problem areas have been identified, small adjustments to the aeration rates in these areas can be made and the effects observed. Once adjustments have been made, the standpipe should be allowed to stabilize for at least a half-hour before making any further adjustments. Aeration rates more than 20% in excess of the calculated theoretical aeration should be avoided as they may result in localized overaeration.

If the standpipe cannot be stabilized by aeration adjustments, the cause of the problem may not be in the standpipe itself. Problems with standpipe entry conditions such as poor hopper design or unstable operation of the fluid bed feeding the standpipe often manifest themselves as problems in the standpipe. When these problems exist, it is often possible to stabilize the standpipe for short periods, sometimes up to several days, by adjusting aeration. Stable operation does not persist, however, and operation will eventually become erratic. Further adjustments to the standpipe aeration will once again stabilize operations for a short time.

This is a strong indication that the condition of the catalyst entering the standpipe is not stable. When this occurs, random changes in the operation of the fluid bed feeding the standpipe or in the flow of catalyst through a poorly designed hopper are affecting the flow of catalyst in the standpipe. After each change, small adjustments to standpipe operation stabilize the catalyst flow, but the next change in the entry conditions produces new upsets in the standpipe.

This type of problem can only be solved by addressing the root cause. Often, the design of the hopper can be improved to minimize the effects of an unstable bed. In some cases, the location of the standpipe entrance must be changed so that the catalyst feeding the standpipe is drawn from a more stable region of the bed.

Gamma Scans and Radiotracer Studies

The use of gamma scans and radiotracer studies has gained widespread acceptance as a useful tool in diagnosing fluidization problems. Gamma scans are used to measure catalyst density in either risers or standpipes. Radiotracer studies are used to evaluate the flow direction and distribution of both catalyst and gas.

Gamma scans are based on the fact that gamma ray absorption is proportional to density.[12] Thus, a standpipe that is full of catalyst will have a higher absorption than an empty pipe.

Diameter scans (Figure 7-11) can be used to detect large bubbles flowing up the standpipe (a sure sign of overaeration). Diameter scans at several elevations can detect large stationary bubbles or catalyst bridges.

Matrix scans (Figure 7-12) are used to measure the density profile of the catalyst. These scans are particularly useful on sloped standpipes since gas bubbles will often collect at the top of the standpipe. Matrix scans can also be used to detect catalyst distribution problems in risers.

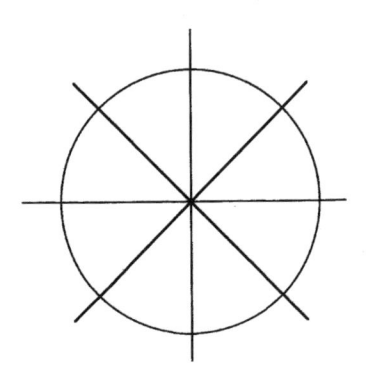

FIGURE 7-11. Gamma diameter scan.

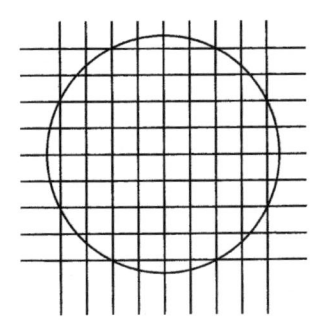

FIGURE 7-12. Matrix scan.

Radiotracer studies involve the injection of a radioactive gas (usually argon 41) or irradiated catalysts and following the movement of the tracer through the system.[12] These tests can be used to determine the direction of gas flow in a standpipe as well as the extent of catalyst/gas mixing in fluid beds. Radiotracer studies have proven to be particularly valuable in identifying problems with catalyst/gas distribution in strippers and regenerators.

Gamma scans and radiotracer studies can be conducted on an operating FCC with minimal interference to normal operation. Since these studies require the use of radioactive materials and specialized equipment, they should be conducted by an experienced contractor who should assume full responsibility for licensing and handling of the radioactive sources.

Cold Flow Models

Cold flow models involve the construction of a scale model of the catalyst system under study. The model is usually constructed of glass or clear plastic to allow for visual observation of the catalyst. Cold flow models can be used for general studies such as the work discussed by Wrench et al.[9] or they can be used to evaluate a specific operational problem and proposed corrective actions. Cold flow modeling is also useful in evaluating the performance of new designs.

Since cold flow modeling requires considerable time and effort in the construction of the model, the cost can be quite high. In addition, specialized models built to address a specific problem often have limited utility for other studies. Thus, this tool is generally reserved for general studies or for the resolution of problems where other approaches have failed to produce a solution.

Slide Valves

The catalyst flow through most FCC catalyst standpipes is controlled by a slide valve or a plug valve. This valve opens or closes as necessary to regulate the flow of catalyst based on the available pressure drop. As discussed in Chapter 2, the pressure

drop across the valve is a function of the unit pressure balance and not the slide valve opening.
The flow of catalyst through the valve can be estimated by:[8]

$$W = 3.57(\Delta P_v \rho_v)^{1/2}$$

where:

W = catalyst mass velocity through valve, lb/ft^2-sec
ΔP_v = pressure drop across valve, lb/ft^2
ρ_v = true density of catalyst/gas mixture, lb/ft^3

In order for there to be a pressure drop through the slide valve, the gas flow must reverse from being upward relative to the catalyst in the standpipe to being downward relative to the catalyst through the slide valve. This reversal often begins several feet upstream of the actual slide valve orifice. This leads to a loss of pressure at the bottom of the standpipe. When this is detected as a part of a standpipe pressure survey, this pressure loss is often a cause of concern. In fact, it is normal and probably indicates that the measured slide valve pressure drop is less than the actual pressure drop. In some cases, relocating the high pressure leg of the slide valve differential meter to measure this additional pressure drop may extend the operating range of the FCC.

References

[1] Letzsch, W.S., "Fluidization and the FCC Process," Katalistiks Fluidization Seminar.

[2] Geldart, D., *Powder Technology,* 7 (1973), p. 285.

[3] Knowlton, T.M., "Evaluating the Effect of Gas and Solid Parameters on the Behaviour of Fluidized Beds of Cracking Catalysts," Katalistiks Fluidization Seminar.

[4] Yerushalmi, J. and Cankurt, N.T., "High Velocity Fluid Beds," *Chemtech,* September (1978), pp. 564–572.

[5] Matsen, J.M., "A Theory of Choking and Entrainment Rates," AIChE 74th Annual Meeting, November (1981).

[6] Baeyens, J. and Geldart, D., Proc., International Symposium on Fluidization and Its Applications (1973).

[7] Abrahamson, A.R. and Geldart, D.,"Behaviour of Gas Fluidized Beds of Fine Powders, Part II, *Powder Technology,* 26 (1980), pp. 35–55.

[8] Knowlton, T.M., "Catalyst Flow in Standpipes," Katalistiks Fluidization Seminar.

[9] Wrench, R.E., Wilson, J.W., and Guglietta, G.,"Design Features for Improved Cat Cracker Operations," First South American Ketjen Catalyst Seminar (1980).

[10] Raterman, M.F., "FCC Catalyst Flow Problems Predictions," *Oil and Gas Journal,* January 7 (1985).

[11] Mott, R.W., "Trouble-Shooting FCC Standpipe Flow Problems," *Davison Catalagram,* No. 83 (1992), pp. 25–35.

[12] *FCCU Tracer Studies,* ICI Tracerco.

✌8
Product Recovery

The product recovery system collects the hydrocarbon vapors from the reactor and separates these into the various gas and liquid products. The primary components of the product recovery system are the reactor transfer line, the main fractionator, and the gas recovery unit (GRU). The liquid products from the main fractionator are typically fractionator bottoms (slurry, decanted oil) and light cycle oil (LCO). Some main fractionators also produce a liquid heavy naphtha product (HCN).

The gas recovery unit—also known as the vapor recovery unit (VRU), gas concentration unit (GCU) or simply as the gas plant—normally produces cat naphtha (WCN, cat gasoline) and olefinic LPG (LPG, olefins) as liquid products. The lightest products are produced as a gas. These are typically referred to as absorber gas, tail gas, dry gas, or fuel gas.

Reactor Transfer Line

The reactor transfer line carries the product vapors from the reactor vessel to the base of the main fractionator. Typically, this line is hot wall alloy with external insulation. In recent years, however, there has been a move to cold wall design for these lines. Cold wall lines are normally carbon steel internally lined with 4–5 inch vibrocast refractory.

Proper design of the transfer line is essential to avoid coke deposition on the line. Coke deposits will restrict the flow area in the line and increase the pressure drop from the reactor to the main fractionator. This will result in lower pressures at the inlet to the wet gas compressor which will in turn limit unit capacity or conversion.

Coke formation in the transfer line is largely due to liquid condensation on the inside surface of the line. The condensed liquids generally remain in the line where they undergo thermal cracking reactions which produce coke deposits. These deposits are usually quite hard and can be difficult to remove from the line.

Heat loss from the transfer line must be minimized to avoid coke formation. The line insulation, whether internal or external, must be in good condition. Areas of damaged or inadequate insulation will almost certainly create cold spots that will promote condensation and coke formation.

Support points are a frequent area of heat loss on externally insulated hot wall lines. The insulation on these points should be given special attention to insure that cold spots will not occur where the supports are attached to the line. Cold wall lines avoid this problem since the insulation is on the inside line surface.

Since some heat loss is inevitable even from well-insulated lines, the length of the transfer line should be minimized. The piping design should avoid large expansion loops as these will add considerable length to the transfer line. Horizontal runs should also be minimized as liquids are likely to collect in these areas.

Expansion joints, purged instrument taps, and manways are all possible sources of heat loss and should be avoided whereever possible. Expansion joints are generally not required, provided the transfer line is properly designed. If manways are necessary in the transfer line, they must be designed to minimize heat loss.

Transfer line design velocities are generally a compromise between the need to minimize pressure drop (thus maximizing wet gas compressor capacity), to insure good vapor distribution in the bottom of the main fractionator and to minimize liquid collection in the transfer line.

Higher transfer line velocities promote the movement of liquid through the line and thus, minimize coke formation. Unfortunately, higher velocities also increase the pressure drop through the line. In addition, high velocities at the inlet to the main fractionator degrade the distribution of the reactor vapors across the bottom section of the main fractionator. This can lead to poor vapor/liquid contact in the bottom of the column.

Transfer line velocities between 100–120 ft/sec are generally a good compromise between the needs to minimize both liquid collection and pressure drop in the transfer line. Lower velocities, between 50 and 70 ft/sec, are usually preferred at the inlet to the main column, however. These conflicting requirements can be addressed by enlarging the pipe diameter before it enters the main tower. This diameter increase should be gradual and should begin 20–30 feet upstream of the main fractionator inlet nozzle.

The reactor transfer line joins the main fractionator at the transfer line flange. This is the location of the transfer line blind that is used to isolate the main fractionator from the reactor during start-up and shut-down. The removal or installation of this blind is probably the single most dangerous activity in the operation of an FCC. The layout of the transfer line flange and associated platforms should assure easy access to the flange and blind as well as reasonable egress for personnel.

Main Fractionator

The purpose of the main fractionator is to cool and desuperheat the vapors from the reactor and to condense the heavier liquid products. This tower is a complex distillation column with several separate sections. The tower itself is divided into desuperheating (or bottoms) section, the heavy cycle oil (HCO) section, the light cycle oil (LCO) section, and the top section. In addition to these column sections, the main fractionator system includes special systems to handle the heavy bottoms product and the overhead condensers and accumulators.

Desuperheating Section

The vapors from the FCC reactor are superheated and must be cooled significantly before products can be condensed. This is accomplished in the bottom section of the main fractionator. This section is often misnamed the flash zone. This terminology is derived from crude distillation units (CDUs) where it is used to describe the bottom section of these towers where hot liquid feed is flashed or vaporized.

In FCC main fractionators, the feed is a superheated vapor. Thus, a more accurate name for the bottom of the tower is the desuperheating zone. In the desuperheating zone, the hot product vapors from the reactor are cooled by contact with a circulating stream of fractionator bottoms product—the bottoms pumparound. In addition to cooling the product vapors and condensing the heaviest liquid product, this circulating liquid stream also serves to wash out any catalyst particles.

Due to the high vapor rates in the bottom of the tower as well as the presence of catalyst dust, conventional trays are not used in this section of the main fractionator. Traditionally, the desuperheating zone has been fitted with several levels of baffle plates, usually shed deck type baffles. The design of these baffles should allow for sufficient pressure drop to distribute the product vapors across the tower cross section, but should also provide considerable open area for vapor flow.

Many FCC main fractionators have been revamped by replacing the existing baffles with structured grid packing. Compared to traditional baffles, properly-installed structured packing provides increased open area for vapor flow while possibly improving vapor liquid contact.

The lower pressure drops associated with structured grid, however, make proper vapor distribution in the bottom of the tower essential. In many cases, these revamps are intended to increase unit capacity. This means that the volume of vapor entering the tower increases following the revamp. This often results in high transfer line velocities and thus, poor vapor distribution in the main column. Many of the problems associated with the use of structural packing can be attributed to high transfer line velocities. These should be checked as a part of the main tower revamp and the transfer line should be enlarged if necessary.

Liquid distribution across the desuperheating section is also critical. Good liquid distribution prevents the formation of dry spots on the baffles or in the grid packing. Dry spots, if formed, lead to high metal temperatures which promote thermal cracking and coke formation in the main column.

Liquid distribution assumes even greater importance with grid packing since there will be little redistribution inside the bed. In addition, once coke has formed in the packing, it is difficult to remove it without damaging the grid.

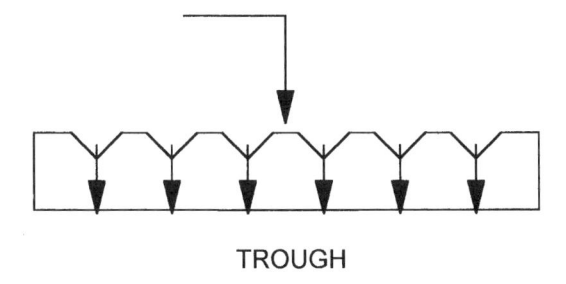

FIGURE 8-1. V-notch distribution.

The circulating bottoms liquid contains catalyst particles and the design of the liquid distributors should take this into consideration. The most effective design is a V-notched trough distributor (Figure 8-1). The circulating rate should be sufficient to fully irrigate the tower internals. As a rule of thumb, the circulation rate should be 1.2–1.5 times the unit feed rate. For packed sections, the liquid rate should be as required by the packing vendor.

The high vapor and liquid loads in the desuperheating zone result in considerable stress on the tower internals in this section. Baffles should be well-braced and secured. Grid packing should be through-bolted. Baffles and distributors should be made of heavier gauge material than used in other fractionation equipment. The lowest baffle, or the bottom of the grid should be no closer than five feet above the vapor inlet.

Most FCC main fractionators have 1 to 2 trays located at the top of the desuperheating zone. These are generally referred to as wash trays. These trays serve to contact the vapors from the desuperheating zone with the internal reflux from the upper section of the tower. They also serve to collect any catalyst that escapes from the desuperheating zone.

The liquid flow over these trays is usually quite low. Traditionally, bubble caps were used in this location to assure that the trays were fully wet. Valve trays should be avoided in this location as they are subject to fouling.

The induced reflux rates to the wash trays are generally very low due to high levels of heat removal from the bottom of the tower. In many towers, the reflux liquid to vapor rate is between 0.10 and 0.20 on a molar basis. Due to this low liquid rate, this section of the tower can easily go dry due to small changes in the tower heat balance.

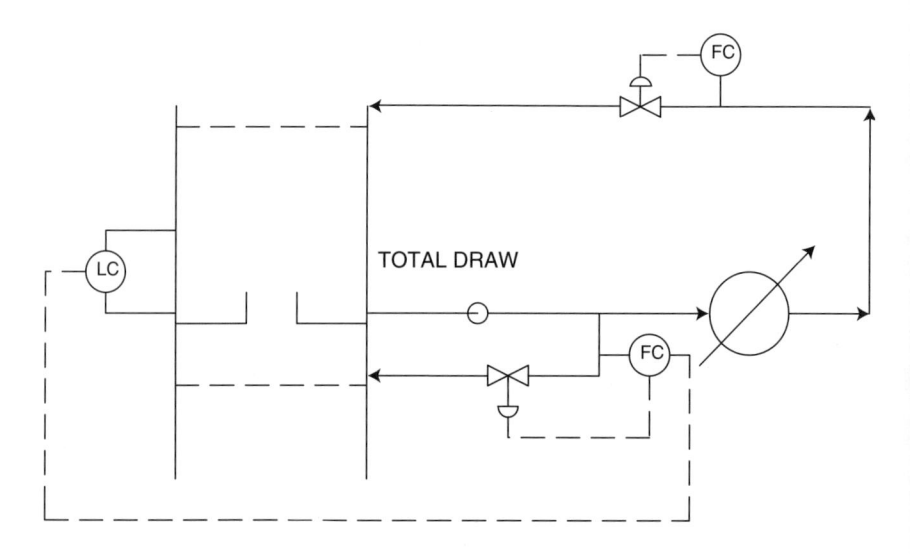

FIGURE 8-2. External heavy cycle oil reflux.

The design shown in Figure 8-2 allows monitoring of the liquid flow to the wash trays. The tower is fitted with a total draw-off pan just above the wash trays. Liquid from this pan is used as the heavy cycle oil (HCO) pumparound and as the reflux to the wash trays. The HCO pumparound is on flow control while the reflux to the bottom of the main column is on flow control reset by the level in the draw-off pan. The liquid rate to the wash trays can now be measured. This allows the operator to note a lack of reflux and adjust tower operation accordingly.

Bottoms Liquid System

The bottoms circulating system is a frequent cause of problems in the main fractionator. The bottoms liquid contains catalyst particles and may also contain coke fragments formed in the main fractionator. In addition, the bottoms liquid contains polynuclear aromatic molecules that can precipitate out of the liquid and deposit on pipes and exchangers. Fouling and plugging are common problems in bottoms systems. When these occur, they generally result in reduced unit capacity, and in extreme cases, may force an early shut-down to clean fouled equipment.

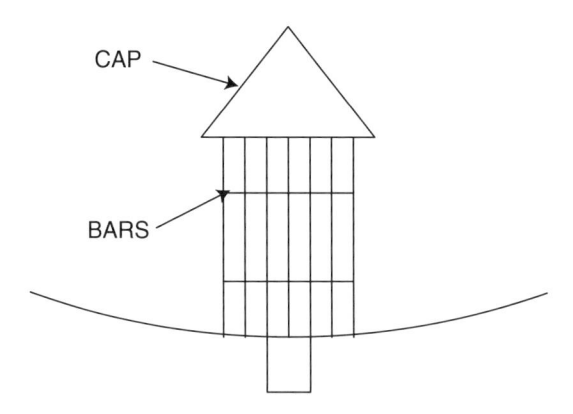

FIGURE 8-3. Main fractionator coke screen.

Bottoms system problems can be reduced by proper design. Complete elimination of these problems may not be possible, however, especially for units processing heavy, aromatic feed stocks.

The bottoms liquid system begins in the bottom of the main fractionator. Here, the net liquid product and the circulating bottoms pumparound collect. The liquid residence time in the bottom of the main column should be minimized as long residence times promote thermal reactions that can lead to the formation of coke or other heavy polynuclear aromatic compounds. If formed, these will precipitate in the tower bottom or at other locations in the bottoms system.

The bottoms of the tower should be cooled using a quench stream of circulating bottoms. Bottoms temperatures should be kept at 650°F or less. The temperature in the desuperheating zone may be higher, provided the bottoms liquid is cooled by the quench. Many towers also are fitted with a steam ring in the bottom head. The purpose of this steam is to provide agitation to the pool of liquid and thus, maintain the catalyst and other solids in suspension.

The bottoms draw-off nozzle should be fitted with a coke screen similar to the type shown in Figure 8-3. The intent of this screen is to catch large pieces of coke and not to screen out catalyst. Smaller screens, in fact, should be avoided as these will usually plug, causing a loss of pump suction.

The bottoms circulating pumps should be located close to the main column, this will minimize the suction line pressure loss and allow for lower liquid levels in the column. Normally, three 50% pumps are installed instead of a single pump and spare.

In recent years, special pumps designed for the hot abrasive slurry found in bottoms systems have come into use. These pumps offer improved resistance to erosion. The hard surface materials used in these pumps are sensitive, however, and can be damaged by rapid heat up. Thus, the spare pump must be kept hot at all times.

The bottoms system piping should be designed with velocities of 6–10 ft/sec. Lower velocities are likely to lead to fouling as catalyst and other solid particles will tend to settle out in the pipes. Higher velocities are generally avoided due to problems with erosion. At least one refinery, however, has reported operation at higher velocities through the use of corrosion resistant pipe[1]. In this case, the erosion problem was actually found to be a corrosion problem aggravated by the continuous removal of the corrosion scale from the piping.

The volume of piping before the heat exchangers should be minimized as the oil in this "hot" volume is subject to thermal cracking. The "hot" residence time in this section is defined as the volume of the liquid in the bottom of the main fractionator plus volume of liquid in the pumps and piping before the first bottoms heat exchanger divided by the net bottoms product rate. Note that the product rate and not the total bottoms circulation rate is used to set the "hot" residence time. Thus, increasing the pumparound rate will not decrease the hot residence time.

The bottoms product should always flow through the tube side of shell and tube heat exchangers. The oil velocity through the tubes should be 6–10 ft/sec. As with the bottoms piping, higher velocities may be possible if corrosion resistant tube materials are used. The tube diameters should be a minimum of 1 inch. Smaller tubes are subject to rapid fouling which reduces the heat transfer as well as increases the tube side pressure drop.

Slurry side fouling is a common problem in main fractionator bottoms systems. Often, even well-designed systems will experience problems with exchanger fouling. In many cases this will cause long-term reductions in feed rate, or at the extreme, force and early shut-down to clean the fouled exchangers.

These problems can be minimized by using parallel exchanger shells in this service. Each shell should be equipped with the necessary valves for complete isolation. This will allow the shell to be bypassed and cleaned on line.

Most systems cool the bottoms liquid by heat exchange with fresh feed in the feed bottoms exchangers and by steam generation in the steam generators. Some refineries also use the fractionator bottoms pumparound to reboil columns in the gas plant. This is generally not a good idea and will often lead to operational problems. The long pipe runs required to bring the fractionator bottoms to the reboiler increase the "hot" volume of the bottoms system. This increases the likelihood of thermal reactions which will produce either coke or a heavy aromatic precipitate.

The return temperature of the bottoms pumparound should be kept above 400°F and preferably at about 50°F. Colder temperatures are possible, but these generally lead to lower circulating rates and thus, poor irrigation of the main column internals. In addition, lower return temperatures encourage the absorption of lighter hydrocarbons into the bottoms liquid and this can lead to low flash points on the bottoms product.

Bottoms Product Filtration

The bottoms product produced from the main fractionator contains catalyst particles. In some cases, these solids will need to be removed before the bottoms can be blended into fuel oil. If the bottoms product is to be used as carbon black oil, solids removal will almost certainly be required. There are three methods used for solids removal: settling, filtration, and electrostatic filtration.

The solids content of the bottoms product can be reduced by allowing the catalyst particles to settle out in a tank. This process can be accelerated by the addition of settling aids. Settling requires no significant changes in unit operation and little, if any, additional equipment. Eventually, however, the catalyst particles must be removed from the tank and disposed of. The catalyst will be saturated with bottoms product which contains a high percentage of polynuclear aromatic compounds. Thus, the catalyst removed from the tanks may need to be treated as a hazardous waste.

The hot slurry product can be filtered using metal filters. Mott offers a filter system fabricated from sintered metal powder, and Pall sells a filter system fabricated from sintered metal mesh. Both of these systems use periodic backwash to remove collected catalyst fines from the filters. This catalyst along with the backwash oil is returned to the riser.

Early installations used HCO as the backwash medium. This led to changes in the feed properties to the FCC during the backwash cycle. Essentially the backwash created an HCO recycle to the riser. Recent installations have eliminated this problem by using fresh feed as the backwash medium.

The Gulftronic Separator uses electostatic filtration to remove catalyst particles from the fractionator bottoms. The fractionator bottoms product flows through modules containing a charged bed of glass beads. Solids in the oil become charged and are collected in the bed.

The modules are cleaned by deactivating and backflushing. As with the physical filters discussed above, the preferred backflush medium is fresh feed.

Heavy Cycle Oil Section

On most main fractionators, the heavy cycle oil section is used only to remove heat from the tower. Some units also draw a heavy cycle oil (HCO) recycle stream which is sent back to the riser. In either case, HCO is not a discrete product. Instead, HCO is a mixture of LCO and fractionator bottoms. The HCO pumparound is withdrawn from just above the wash trays (preferably from a total draw of pan as discussed above). This pumparound is generally used to preheat fresh feed and to reboil the debutanizer tower in the gas plant. It may also be used to generate medium pressure steam and to preheat boiler feed water. The cooled pumparound is returned to the top of the HCO section.

Fractionation between the fractionator bottoms and the light cycle oil product is generally not important. Thus, most FCC main fractionators are designed to maximize the combined heat removal from the HCO and bottoms pumparound. As discussed previously, the internal reflux to the wash trays is generally set at a minimum value. This sets the bottoms pumparound duty. The HCO pumparound duty is generally set so that the internal reflux above the pumparound return is equal to 0.3–0.5 times the vapor leaving the top of the HCO section on a molar basis.

Light Cycle Oil Section

Light cycle oil (LCO) is generally the heaviest side draw from the main column. Since fractionation between LCO and HCO is not generally an important consideration, the number of trays between the LCO draw and the HCO pumparound return is small. The light cycle oil product is stripped in a steam stripper to control the flash point. The stripper normally has 4–6 trays. Overhead vapors from the stripper are returned to the main column.

The light cycle oil pumparound is used to reboil towers in the gas plant (usually the stripper) and to preheat fresh feed. Unstripped light cycle oil is also frequently used as sponge oil for the secondary (sponge) absorber. The rich sponge oil from the secondary absorber is returned to the main column along with the LCO pumparound return.

Separation between the LCO and naphtha products is an important design and operating parameter in FCC main fractionators. Typically, the tower is designed to give a gap of 30–50°F between the ASTM 5% point of the LCO and the ASTM 95% point of the naphtha stream. This results in a fairly high internal reflux rate above the LCO section. Typically, the internal reflux ratio in this section of the tower is 0.5–0.7 on a molar basis.

Some main fractionator designs use a heavy naphtha pumparound in place of the LCO pumparound. The draw-off for this pumparound stream is located several trays above the LCO draw. These designs generally use heavy naphtha for sponge oil in the secondary absorber. Heavy naphtha is less likely than LCO to cause foaming in the sponge absorber. It is, however, less effective as an absorbent and a higher sponge oil rate is generally required.

Top Section

There are two basic designs used for the top section of the main fractionator. In one design, the top of the tower is cooled using a top pumparound. Heat removed from this stream can be used to preheat fresh feed and/or boiler feed water. If the gas plant includes a depropanizer (C_3/C_4 splitter) it may also be possible to use this stream to reboil this tower. Generally, there is also a small reflux stream from the overhead condenser. This stream is used to control the top temperature of the tower.

The other system uses a larger reflux stream from the overhead condenser. In this system, the heat removed from the top of the main fractionator is rejected to air and water coolers and not recovered.

The addition of a top pumparound to the main fractionator increases the overall energy recovery of the system (provided the heat removed is not simply rejected to cooling water or air). The top pumparound system does, however, add additional cost and thus, should only be used if justified by the cost of the recovered energy.

In some FCCs, a heavy naphtha side product is drawn from the top section of the main column. This product is generally stripped in a steam stripper to control flash point. The overhead temperature of the main column should be above the condensation point for the water in the vapor stream leaving the tower. This sets a lower limit on the cut point between any heavy naphtha product side draw, and the light naphtha taken overhead. If water is allowed to condense in the top of the column, it can lead to corrosion and fouling of the upper column trays. The buildup of water in the top of the tower will also result in poor fractionation and possibly cause early flooding of the top section.

Overhead System

Vapors from the top of the tower are partially condensed in the overhead condenser system. Generally, the main overhead condensers are air coolers. These are followed by water-cooled trim condensers. The vapor/liquid mix from the condensers flows to the overhead separator. Vapors from this drum are sent to the wet gas compressor. This wet gas stream contains most of the C_4 and lighter products produced in the FCC. A portion of the liquid from the overhead separator is returned to the main column as reflux. The net liquid product, unstabilized naphtha, is sent to the gas plant.

A common problem in main fractionator overhead systems is corrosion and fouling of the overhead condensers. This generally occurs at the point where water first begins to condense from the tower overhead vapors. This water contains a high concentration of hydrogen sulfide and other corrosive agents. The problem can be solved by introducing a water wash upstream of

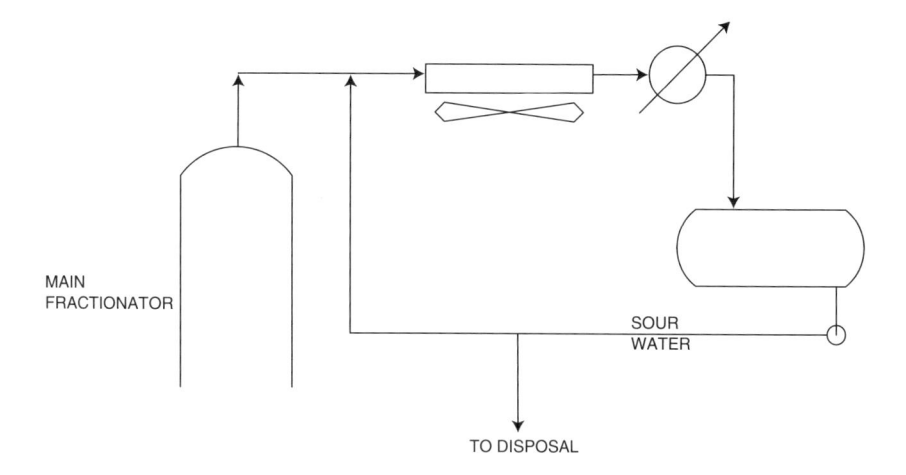

FIGURE 8-4. Main fractionator overhead water wash.

the condenser (Figure 8-4). Normally this is done by recycling a portion of the water collected in the overhead separator. The injection rate is set to insure that 5–10% of the injected water remains as a liquid.

If water is recycled from the overhead condenser, there is no net effect on the condenser duty. The temperature of the gas entering the condensers is, however, lower. This reduces driving force for heat exchange in the condensers and may increase the required surface area. Thus, the capacity of the condensers should be checked before retrofitting this type of wash system on an existing unit.

Tray Efficiencies

Tray efficiencies vary across the main fractionator. The baffled section of the desuperheating zone is primarily a heat transfer section. Typically this zone will provide between one and two theoretical stages of separation.

Tray efficiencies from the wash trays to the light cycle oil draw are generally low, about 50%. Pumparound sections can generally be considered to be one stage. Above the LCO draw, the tray efficiencies increase to about 70%. As before, pumparound sections should be considered to be one stage.

Gas Recovery Unit

The gas recovery unit (GRU) processes the gasoline and lighter materials and separates these into the desired product streams. The primary components of the GRU are the wet gas compressor, the absorber/stripper, the secondary absorber, and the debutanizer. Depending on the product slate, the GRU may also include a depropanizer and/or a gasoline splitter.

Most gas recovery units also include facilities for treating the various product streams. These may include amine treating for the removal of hydrogen sulfide as well as facilities for removal or conversion of mercaptans.

Wet Gas Compressor

The low pressure gas stream from the main fractionator overhead system contains most of the C_4 and lighter products produced in the FCC reactor. This stream also contains any inerts carried over from the regenerator. It is also saturated with water. The wet gas compressor raises the pressure of this gas stream so that the C_3 and C_4 hydrocarbons can be recovered in the GRU.

Compressor Types. Modern wet gas compressors are two-stage centrifugal machines. Older FCCUs, however, were built with reciprocating wet gas compressors (also two-stage). Many of these older machines are still in use. These old reciprocating compressor systems are less reliable than modern centrifugal machines. The cost of a new compressor is, however, high and it is difficult to justify replacement on the basis of improved reliability.

Reciprocating compressor systems consist of several separate two-stage compressors operating in parallel. Since the reliability of reciprocating compressors is low, one compressor is often held in reserve as a spare. This allows for rapid recovery to normal operation after the failure of an operating compressor.

In a modern compressor, both stages are on a common shaft and are housed in a single casing. The compression ratios of each stage are set to balance the power requirements.

Gas from the first stage is cooled in an interstage condenser and the condensed liquids are separated from the gas before it

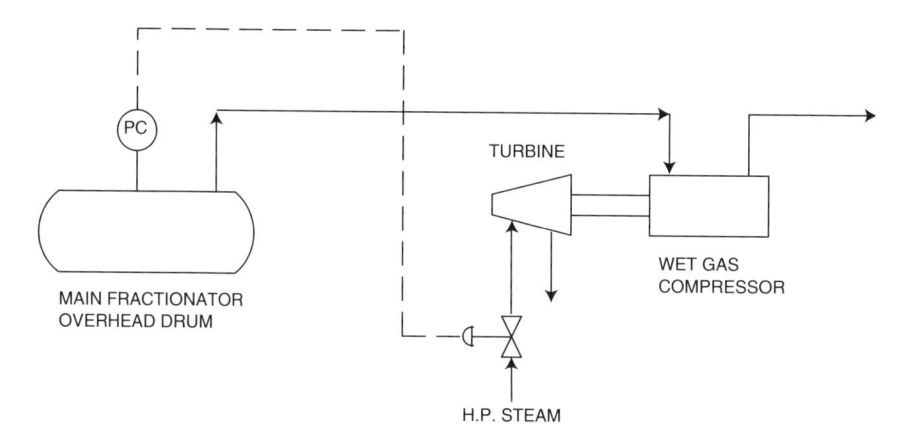

FIGURE 8-5. Main fractionator pressure control.

flows to the second stage. The second stage discharge pressure determines the level of propylene recovery possible in the GRU. Modern FCCs are generally designed for maximum recovery. On these units, the discharge pressure is usually 220–250 psig. Many older units were designed for lower recoveries. Discharge pressures on these units were generally in the 160–175 psig range.

Drivers. Centrifugal wet gas compressors generally use steam turbine drivers. On these systems, the turbine speed is varied to control the compressor suction pressure (Figure 8-5). This is a primary control point on the FCC as the pressure in both the main fractionator and the FCC reactor float on compressor suction pressure. The combination of a centrifugal compressor with a variable speed turbine drive provides responsive pressure control and is superior to systems using constant speed compressors and pressure control valves. A few systems have also been built using variable speed electric drives.

Reciprocating compressors are generally driven using electric motors or, on some early systems, diesel engines. Suction pressure (and thus reactor pressure) is controlled by the use of valves in the compressor inlet and by recycling gas from the compressor discharge.

Water Wash. The wet gas from the main fractionator contains various corrosive compounds produced in the FCC reactor. Chief among these is hydrogen sulfide. Thirty to fifty percent of the sulfur in the FCC feed leaves the reactor as H_2S. In addition, some of the nitrogen compounds in the feed are converted to hydrogen cyanide and ammonia. The overhead vapors from the main fractionator contains water. This water is condensed at various points in the wet gas system.

In the wet gas system the hydrogen sulfide can react with iron to form iron sulfide:

$$2H_2S \rightarrow 2H^+ + 2HS^-$$
$$2HS^- + Fe \rightarrow FeS + S^{-2} + 2H^o$$

In the absence of cyanide, this reaction does not present a major problem since the iron sulfide formed will actually form a protective scale on the metal surface. In addition, the monatomic hydrogen will combine to form $H_2{}^2$.

Cyanide will, however, react with the iron sulfide scale to form ferrocyanides which are soluble in the condensed water.

$$FeS + 6(CN)^- \rightarrow Fe(CN)_6 + S^{-2}$$

Thus, the protective scale is removed and the metal surface is exposed to the corrosive environment. In addition, the presence of cyanide inhibits the formation of H_2. The monatomic hydrogen will then diffuse into the steel where it will cause hydrogen blistering.[2]

The most commonly used method to control corrosion in the wet gas system is a continuous water wash of the compressor discharge streams. This water wash acts to dilute the corrosive components in the aqueous phase and thus, reduces their corrosive activity. Water wash systems can use cocurrent, countercurrent, or parallel injection systems.

In the cocurrent system, fresh wash water is injected into the first stage compressor discharge. This water along with water condensed in the interstage condenser is collected in the interstage separator drum. A portion of this collected water is then pumped into the second stage compressor discharge (Figure 8-6). The water collected in the high pressure separator is combined with the water from the main fractionator overhead drum and sent to the sour water system.

Product Recovery **239**

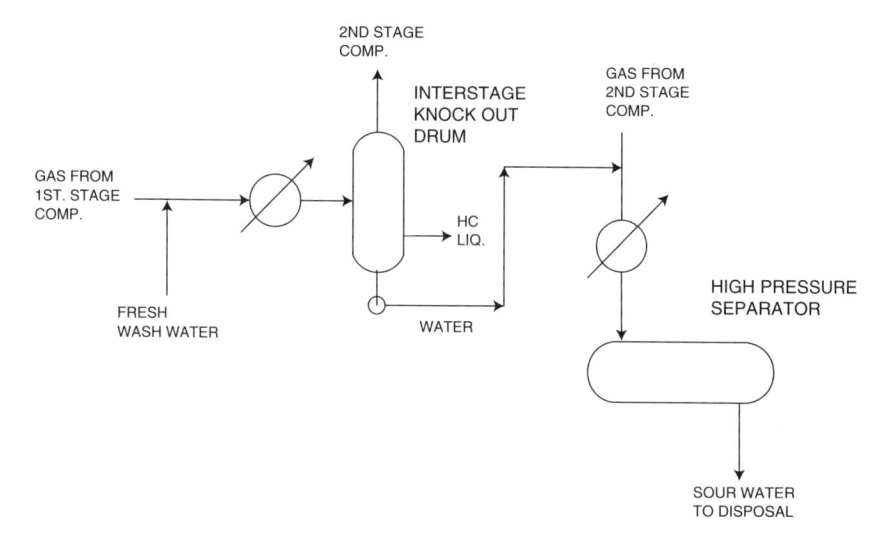

FIGURE 8-6. Cocurrent water wash.

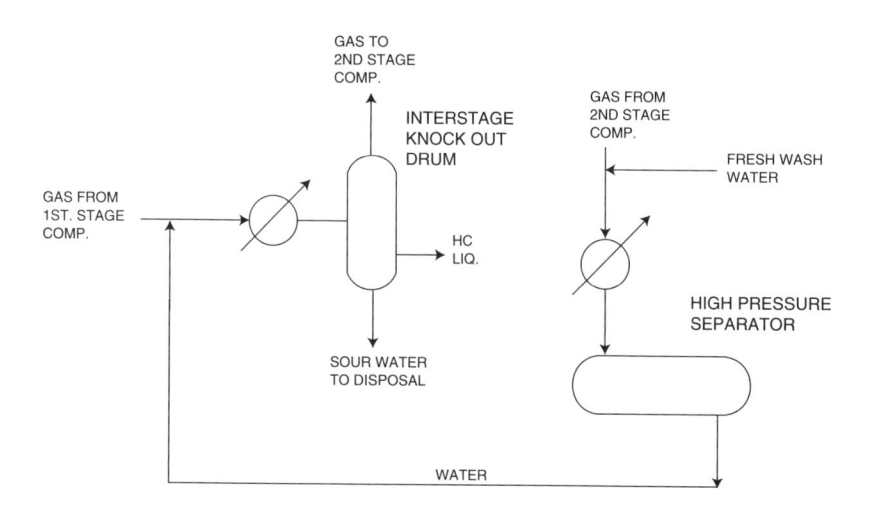

FIGURE 8-7. Counter current wash.

In the countercurrent system, the fresh wash water is injected into the second stage discharge. This water is then collected in the high pressure separator and injected into the first stage compressor discharge (Figure 8-7). Water from the interstage separator drum is sent to the sour water system.

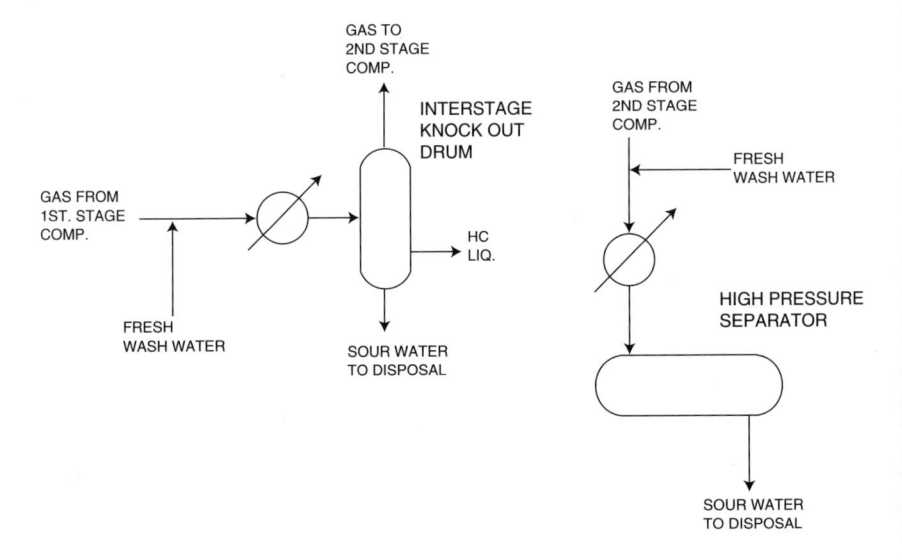

FIGURE 8-8. Parallel wash.

In the parallel system, fresh wash water is injected into both the first stage and second stage compressor discharges. Water from both the interstage drum and the high pressure separator is sent to sour water (Figure 8-8). Regardless of the injection scheme, the injection rate to each stage is set at 1–2 gallons per minute for every 1000 barrels of FCC capacity.

Parallel injection systems provide the best dilution. Since fresh water is used in each stage, there is no chance that captured materials will be released during a subsequent injection. Parallel systems do, however, require twice the wash water and this greatly increases the overall sour water produced in the FCC.

Countercurrent systems are the least effective. Some of the corrosive materials collected in the high pressure stage will be revaporized when the water is reinjected into the lower pressure first stage discharge. These materials will then be recycled back to compressor second stage. The net result is a higher concentration of corrosive compounds in areas of highest pressure. Thus, cocurrent injection should be used where wash water or sour water limits prevent parallel wash water injection.

Various chemicals have been used in combination with water wash systems to reduce corrosion. Polysulfides react with hydrogen cyanide to form thiocyanate. Thiocyanates do not react with iron sulfide and thus, the protective layer of sulfide scale is preserved. Filming inhibitors coat the inside metal surface to prevent contact with the aqueous phase. Passivators interact with the surface to inhibit corrosion.[2]

Absorber/Stripper

The absorber stripper system is the primary liquid recovery section in the gas plant. This system consists of the primary absorber and the stripper. In a modern gas plant these are two separate towers that are linked together through the high pressure separator.

Primary Absorber. The gas from the second stage compressor discharge is mixed with liquid leaving the bottom of the primary absorber, liquid from the compressor interstage and vapors from the overhead of the stripper. This combined steam is cooled in the high pressure condenser and then sent to the high pressure separator. This cooled contact serves as the first absorption stage for the recovery of C_3 and C_4 LPG from the gas steam.

The high pressure condenser is usually an air-cooled exchanger followed by a water-cooled trim condenser. Considerable water is condensed at this point. This water along with wash water injected into the second stage compressor discharge stream must be removed from the high pressure separator. Gas from the high pressure separator is fed to the primary absorber.

The primary absorber recovers most of the C_3 and C_4 LPG. Modern gas recovery units can recover between 90 and 95% of the propane and propylene produced in the FCC. Butane/butylene recovery generally exceeds 98%. In older, lower pressure GCU's propylene recovery may be as low as 85%.

Low pressure distillate from the main fractionator overhead drum is pumped to the top of the primary absorber where is serves as lean oil. Additional lean oil may be added by recycling debutanizer bottoms (stabilized cat naphtha) to the top of the absorber. This recycled lean oil may be introduced into the tower at the same location as the main fractionator overhead liquid, or it may be added one or two trays above the liquid from the main fractionator.

Some designs incorporate an overhead contact (presaturater) drum as a part of the primary absorber. In these designs, the recycle lean oil is mixed with the vapors leaving the primary absorber. This combined stream is cooled and sent to a separation vessel. Liquid from this vessel is sent to the top of the primary absorber while the vapors flow to the sponge absorber.[3]

Regardless of the design configuration, the use of recycle lean oil increases the overall liquid load in the primary absorber. It also increases the liquid flow to the high pressure condenser and the high pressure separator. In addition, recycle lean oil increases the liquid load and the reboiler duty in the stripper tower and the debutanizer. The use of recycle lean oil can then, affect the capacity of the gas recovery units.

The optimum rate of recycle lean oil must be determined based on the trade-off between the increased recovery achieved versus possible losses in capacity or increases in energy use. This can best be done by developing a computer model of the gas concentration unit using a flow sheet simulator.

The primary absorber usually contains between 20 and 30 trays. Tray efficiencies in this tower are low, about 33%. Thus most primary absorbers have between 7 and 10 theoretical stages. Additional stages cannot generally be justified. Additional stages produce only a small increase in recovery. This increased recovery is usually not sufficient to offset the increased cost associated with the additional stages.

The absorption of light hydrocarbons into the lean oil is exothermic, thus heat is released in the primary absorber. This, coupled with the cooling effects of the liquid and vapor feed streams produces higher temperatures in the middle of the absorber.

The high temperatures in the center of the absorber reduce recovery. This is especially true for absorbers operating with limited lean oil rates. On these towers, recovery can be improved by installing an intermediate cooler as shown in Figure 8-9. Liquid flowing down the tower is withdrawn using a total draw-off tray. This liquid is passed through a cooler and returned to the absorber on the tray below the draw-off.

The installation of the intermediate cooler has little effect on the temperatures at the top and bottom of the absorber tower. Instead, this cooler reduces the temperature in the middle of the tower and this results in a more even temperature profile through the column.

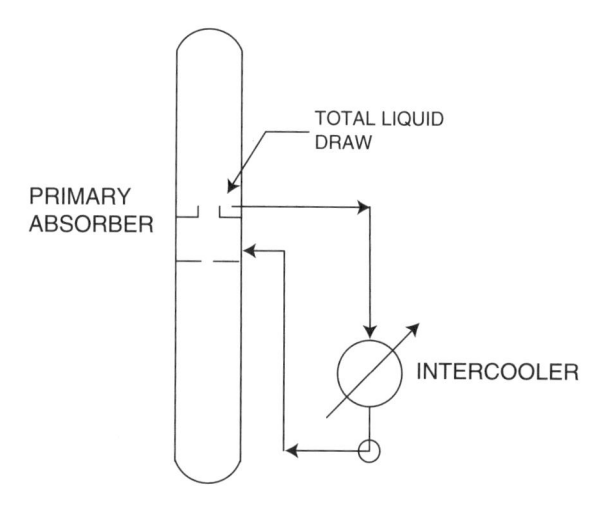

FIGURE 8-9. Primary absorber intercooler.

Stripper. The liquid from the primary absorber contains the recovered C_3 and C_4 LPG. It also contains C_2 and lighter hydrocarbons and hydrogen sulfide recovered in the absorber. Most of these undesirable species are removed in the stripper.

The primary absorber bottoms liquid is initially mixed with the stripper overhead vapors and the vapors leaving the second stage of the wet gas compressor. This combined stream is cooled and sent to the high pressure separator.

Liquid from the high pressure separator is sent to the top tray in the stripper. The stripper tower is reboiled, usually by the heavy pumparound from the main fractionator. The reboiler duty may be controlled either by the temperature at the bottom of the stripper or by the overhead vapor rate.

The stripper is intended to remove the C_2 and lighter gases from the recovered liquid products. If these are not removed they will be carried to the debutanizer where they will collect in the overhead system as uncondensable gases. A typical design specification is a maximum of 2.0% ethane in the C_3/C_4 LPG product.

The stripper tower normally contains 20–30 trays and tray efficiencies are approximately 33%. As with the primary absorber, the additional trays are usually not cost effective. In many units, the primary absorber and stripper are stacked with the stripper on the bottom.

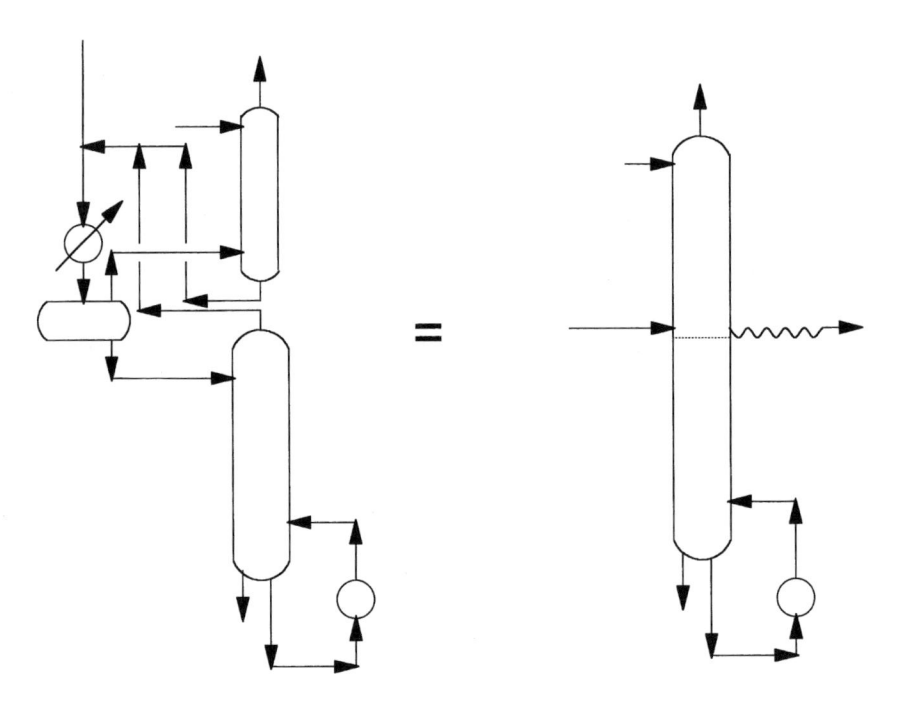

FIGURE 8-10. Absorber/stripper as a single tower.

The high pressure condenser, high pressure separator, primary absorber and stripper can actually be thought of as a single distillation column (Figure 8-10). This approach simplifies computer simulation of this system since it eliminates several recycle streams. Some older GRUs in fact, were designed without a high pressure condenser/separator before the absorber stripper tower (Figure 8-11). In these units, the absorber/stripper system is often called the deethanizer.

Water Draw. Water is removed from the high pressure separator. Even so, water can often find its way into the absorber/stripper tower. Once inside this tower, water cannot escape and will tend to accumulate over time. This can be addressed through the addition of a water draw off to the stripper tower.

Figure 8-12 shows a typical water draw-off design. The draw off consists of a collection pan located in the tower. This pan need only be large enough to divert the water/hydrocarbon mixture into

Product Recovery **245**

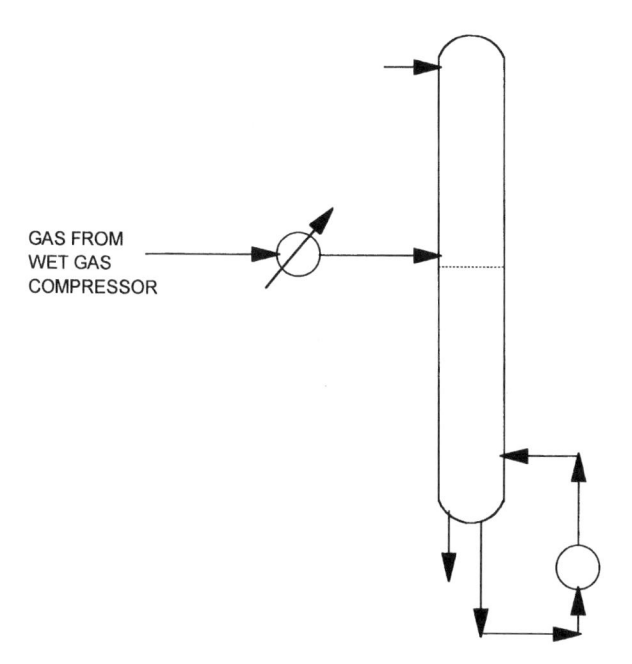

FIGURE 8-11. De-ethanizer.

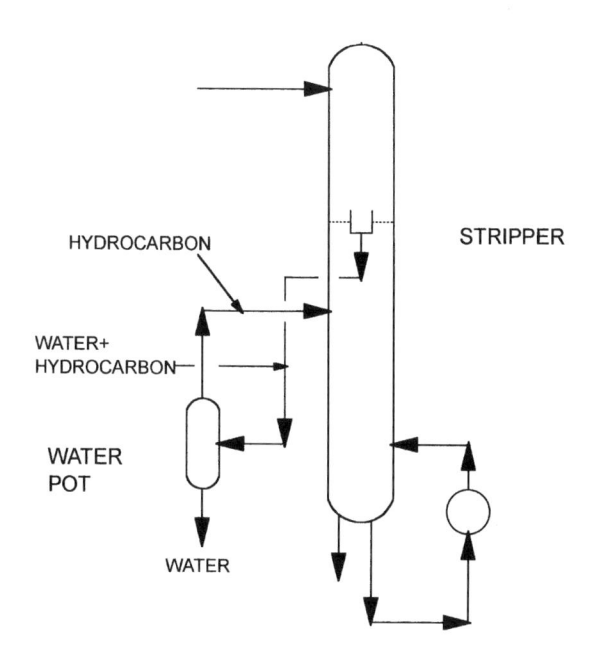

FIGURE 8-12. Stripper water draw off.

the draw-off line.[4] The draw-off line goes to a separation pot. This small vessel is normally located at grade. The separation of the water and hydrocarbon phases takes place in the pot. Hydrocarbon from the top of the pot flows back to the stripper and is introduced into the tower below the draw-off pan. Water collected in the pot is drained off periodically. The flow through the system is driven by the density difference between the water/hydrocarbon mixture in the draw-off leg and the hydrocarbon in the return leg. When there is no water present in the stripper, the net flow is zero.

Secondary Absorber

The gas leaving the top of the primary absorber contains C_5 and heavier hydrocarbons that are lost from the lean oil. These lost lean oil components represent lost product. In addition, if not recovered, they can cause problems in the refinery fuel gas system. The secondary or sponge absorber is intended to collect these heavy hydrocarbons.

Lean oil to the sponge absorber is either unstripped light cycle oil or unstripped heavy naphtha from the main fractionator. Light cycle oil is a more effective absorbent, but is more likely to cause foaming in the sponge absorber. The lean oil from the main fractionator is cooled first by exchange with the rich oil from the bottom of the sponge absorber and finally by air and/or water coolers. The rich oil, after being heated by exchange with the lean oil, is returned to the main fractionator.

In addition to the lean oil components lost from the primary absorber, the rich oil will contain some absorbed lighter gases such as C_3 and C_4 hydrocarbons. These, along with the C_5 and heavier materials collected from the gas stream are flashed off in the main column and returned to the GRU via main fractionator overhead system and the wet gas compressor.

It is a common misconception, even among experienced FCC engineers, that the sponge absorber is intended to increase the recovery of C_3 and C_4 hydrocarbons. In fact, as discussed above, the sponge absorber is intended to collect gasoline range material lost from the primary absorber lean oil. Increasing the lean oil rate to the sponge absorber may produce a small increase in the recovery of lighter hydrocarbons. It will also increase the volume of recycle gas that must be handled by the wet gas compressor. This can result in reduced capacity, or in some cases, reduced overall recoveries.

Debutanizer

The stripper bottoms contains the total recovered liquid product from the GRU. The debutanizer—or naphtha stabilizer—is designed to split this liquid into a naphtha product and a mixed LPG product.

The debutanizer is a classic distillation column. The feed liquid enters in the middle of the tower. The lighter LPG product is produced from the overhead system and the heavier naphtha product is produced from the bottom of the tower. This tower is normally reboiled by the heavy cycle oil pumparound from the main fractionator.

The debutanizer normally operates at a lower pressure than the stripper and the stripper bottoms flows to the debutanizer by pressure. Normally, this stream is preheated by contact with the stabilized naphtha product before being fed to the tower. The feed generally flashes as it passes through the control valve used to regulate flow. Thus, the feed enters the tower in two-phase flow. Before the development of computer flow sheet simulators, this made the location of the optimum feed tray difficult. Some older debutanizer designs actually have multiple feed locations to allow for field optimization of the feed tray.

In modern designs, the optimum feed tray location should be selected during the process design phase based on the feed composition and conditions.

The tower overhead is totally condensed and subcooled in the overhead condensers. Most debutanizers use a hot vapor bypass to control the overhead drum pressure (Figure 8-13). In these towers, there is always a vapor space above the liquid in the overhead drum. If, due to poor stripping, the feed to the debutanizer contains excessive ethane and lighter hydrocarbons, these will collect in this vapor space. Eventually, these gases must be vented to prevent high pressures in the drum. This vent may go to either the flare or to the fuel gas system. In either case, valuable product

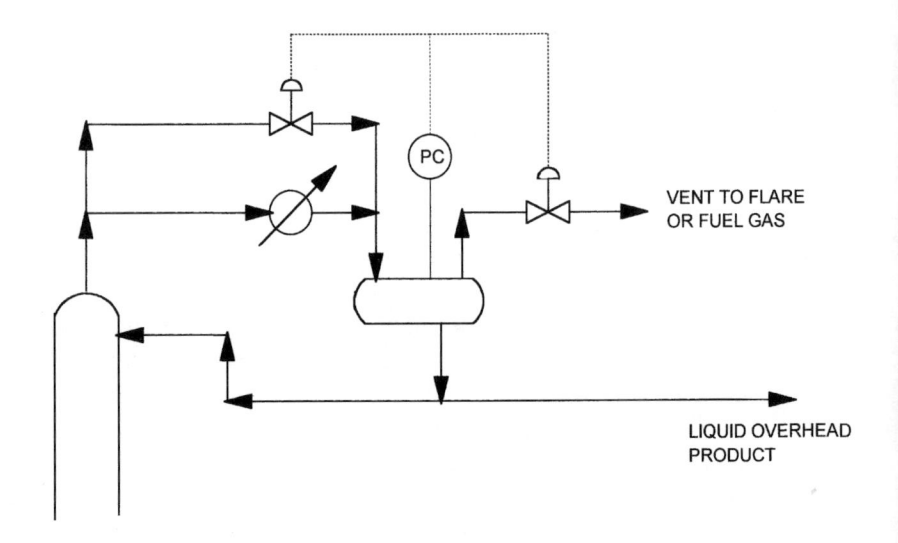

FIGURE 8-13. Hot vapor bypass.

will be lost since the vent gas will contain a high concentration of C_3s and C_4s.

The bottoms product from the debutanizer is the net gasoline product from the FCC. This stream is cooled, usually be exchange with the tower feed and then by air and/or water cooling, before being sent to treating and product blending.

Other Towers

Many gas recovery units contain additional distillation columns to further separate the liquid products. The C_3/C_4 LPG from the debutanizer can be split into separate C_3 and C_4 product in a depropanizer. This tower is usually reboiled using steam. As with the debutanizer, the tower overhead is totally condensed and subcooled in the overhead condenser.

The gasoline product from the debutanizer can be split into light cat naphtha and heavy cat naphtha in a gasoline splitter.

Approximately 70% of the C_3 LPG produced in the FCC consists of propylene. A propane/propylene splitter can be used to produce either chemical grade (normally about 95%) propylene or polymer grade propylene (normally 99+% purity). Where a market exists, these products are often the most valuable produced from the catalytic cracker. Polymer grade propylene especially, can be a very valuable product.

The propane/propylene product from the depropanizer contains all of the ethane and lighter materials that were not removed in the stripper. The quantity of these components is usually in excess of the allowable specification for polymer grade propylene. Thus, the C_3 stream must first be deethanized.

The deethanized C_3s are then sent to the propane/propylene splitter. This tower generally holds 150–200 trays. Since the boiling points of propane and propylene are very close, there is little temperature difference between the top and the bottom of the propane/propelyene splitter. In addition, the high reflux rates required in this separation result in a large volume of internal traffic compared to the feed and product rates. These facts make control of the tower by traditional temperature controls very difficult. Advance control systems based on product and feed rates and on line product analysis are usually required.

Since the temperature difference across the tower is so small, a heat pump can be used to transfer heat from the overhead condenser to the reboiler. This will result in considerable energy savings on this tower.

Treating

The gas and liquid products from the GRU contain sulfur compounds, chiefly hydrogen sulfide and various mercaptans. These must generally be removed or converted to sweet sulfur compounds for the products to meet specifications. The treating units used for this are often contained as a part of the GRU.

Fuel Gas Treating

The fuel gas leaving the sponge absorber contains hydrogen sulfide. It also contains carbon dioxide carried to the reactor by the catalyst leaving the regenerator. These two contaminants are removed in the fuel gas amine absorber.

The fuel gas first passes through a knock-out drum to remove any liquid hydrocarbon carried over from the sponge absorber. If liquid hydrocarbon were to enter the amine absorber, it would cause foaming of the amine solution and contamination of the fuel gas system.

The gas from the knock-out drum enters the bottom of the amine absorber. Cold lean amine from the amine regenerator enters the top of the tower. To avoid hydrocarbon condensation and foaming, the lean amine should always be a minimum of 10°F hotter than the gas entering the tower.

The absorber tower can be either a trayed or a packed column. In either case, there should be a water wash tray followed by a mist eliminator located at the top of the tower (Figure 8-14). The water wash tray serves to collect any amine solution entrained into the sweet fuel gas. The mist eliminator collects any water/amine mist leaving the wash tray.

LPG Amine Absorber

The C_3/C_4 LPG also contains hydrogen sulfide. This is removed in the LPG amine absorber. This is normally done before the LPG is split if the GRU includes a depropanizer.

The LPG amine absorber is generally a packed tower, but trayed columns have been used. The tower is liquid full with the

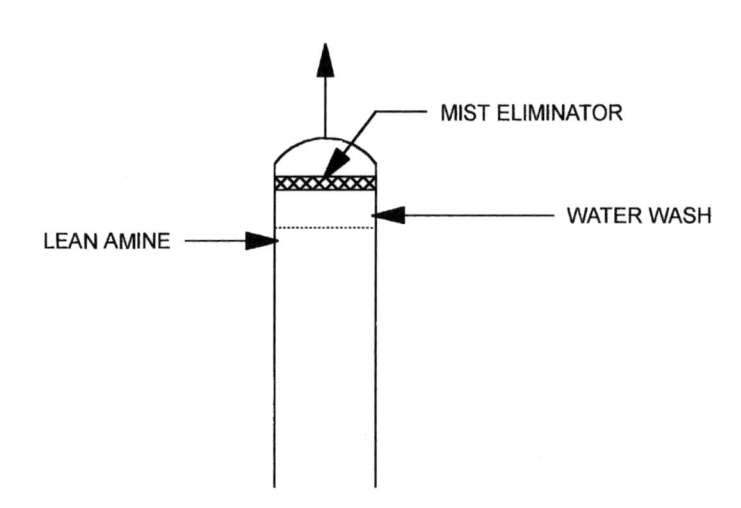

FIGURE 8-14. Fuel gas amine absorber water wash tray.

amine forming the continuous phase. The LPG is introduced at the base of the column and flows upward through the amine phase. Fresh amine is added at the top of the column. The interface between the aqueous amine phase and the LPG is generally held at the top of the column.

Amine Regenerator

The rich amine from both the fuel gas and LPG absorbers is regenerated in the amine regenerator. In many refineries, this tower is not located within the FCC battery limits. In others, the amine regenerator is included as a part of the gas concentration unit.

The rich amine from the absorbers flows first to a hydrocarbon release drum where volatile hydrocarbons are flashed off. The rich amine then flows to the regeneration tower. Typically, the amine enters two trays below the top of the tower.

The regenerator is reboiled using low pressure steam. High tube metal temperatures in the reboiler can lead to rapid corrosion in this exchanger. To prevent this problem, the steam temperature should be no more than 250°F.

Overhead vapors from the regenerator are partially condensed. The gas from the condenser contains the hydrogen sulfide and carbon dioxide absorbed by the amine. The liquid, mostly water, is returned to the tower as reflux. A small purge stream of water is removed to balance the amine concentration.

Lean amine is drawn from the bottom of the tower and cooled by exchange with the rich amine feed and then by air and/or water coolers. A slip stream of lean amine is filtered to remove solids; in some cases, this stream also flows through activated carbon beds to remove dissolved hydrocarbons.

Mercaptan Extraction

The C_3/C_4 LPG contains mercaptan (R – SH) compounds that must be removed. This is done through extraction with a caustic solution. A typical mercaptan extraction process is shown in Figure 8-15.

The LPG first passes through a caustic prewash to neutralize any residual hydrogen sulfide not removed by amine absorption.

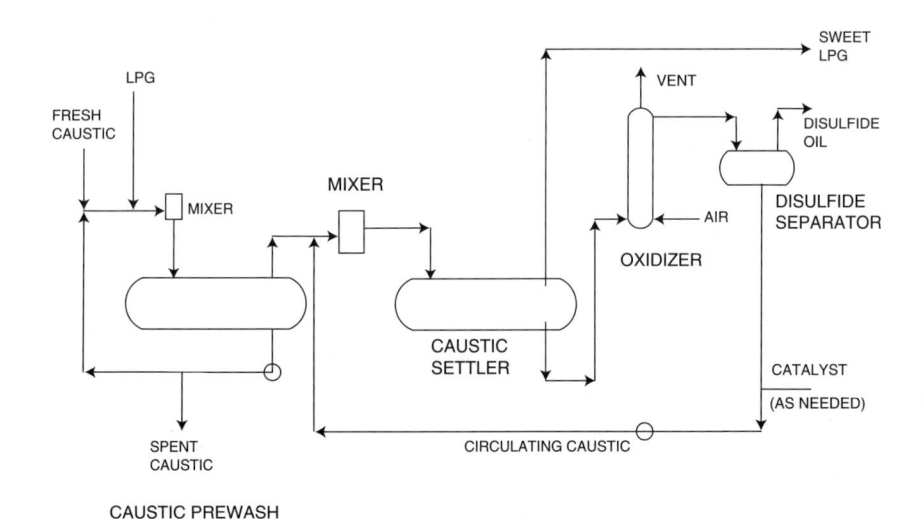

FIGURE 8-15. LPG extraction.

From this prewash, the LPG flows to the extraction stage. Here, the mercaptans react with a caustic solution:

$$RSH + NaOH \rightarrow RSNa + Na + H_2O$$

The treated LPG flows to storage or other use. The aqueous caustic solution flows to the oxidizer.

In the oxidizer, the mercaptans are oxidized to disulfide oils:

$$2RSNa + 1/2\,O_2 + H_2O \rightarrow RSSR + 2NaOH$$

This reaction is accelerated by the presence of an oxidation catalyst in the caustic solution. The disulfide oil is insoluble in the aqueous caustic solution and is removed in the disulfide separator. The caustic is returned to the extraction vessel. The disulfide oil is usually mixed with refinery fuel oil or burned in a refinery furnace.

Gasoline Sweetening

The gasoline products produced in the FCC also contain mercaptan compounds. These compounds are corrosive and must be

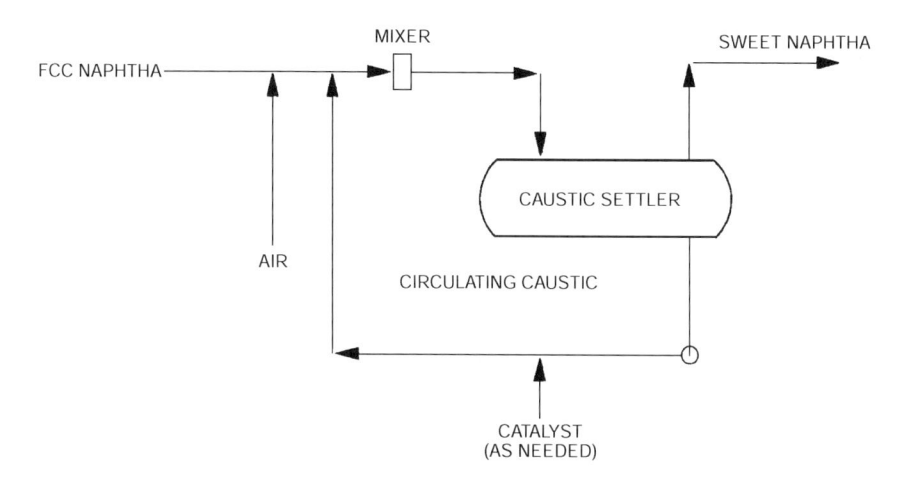

FIGURE 8-16. Wet sweetening process.

converted to a less harmful form. Normally, however, they are not removed from the gasoline stream.

Gasoline sweetening converts the mercaptan compounds to disulfides. The chemical reactions are essentially the same as those used in LPG treating. In gasoline treating, however, the disulfides formed are reabsorbed into the hydrocarbon stream.

Wet processes using a caustic solution can be used for gasoline sweetening. A typical sweetening unit is shown in Figure 8-16. The most widely used process, however is the Minalk process licensed by UOP. This process (Figure 8-17) uses activated carbon beds which support the oxidation catalyst. Gasoline caustic and air flow over these beds where the mercaptans are converted to disulfides.

Operating Considerations

Operation of the product recovery section can affect product quality, unit capacity and unit safety. Each of the various parts of this section involves specific design and operating considerations that should be well understood by responsible engineers.

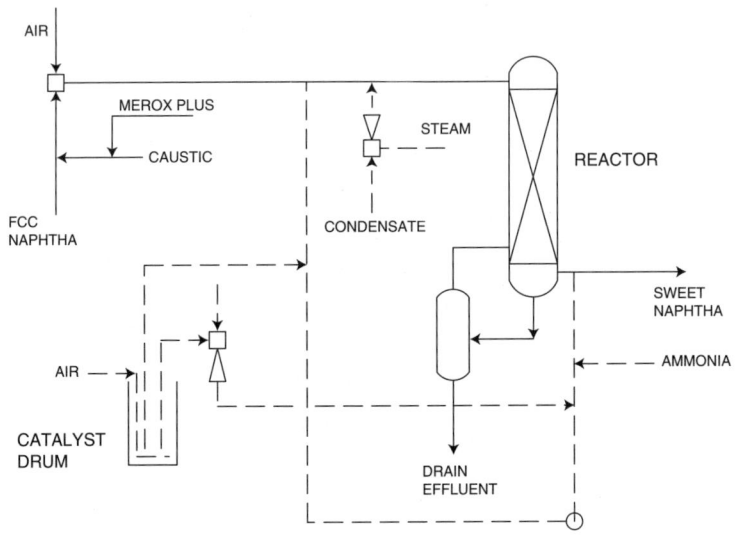

FIGURE 8-17. UOP Minalk sweetener. (Drawn with permission from information supplied by UOP. Merox and Minalk are trademarks of UOP.)

Reactor Transfer Line

The primary concern with the reactor transfer line involves the removal of the transfer line blind during start-up. This blind is usually quite large, and its removal exposes personnel to steam and possibly air at elevated temperatures. There is also a risk of fire, as hydrocarbons are usually present in the main fractionator at the time of blind removal.

The best approach to this step is to remove the blind after the catalyst has been loaded into regenerator and a steam purge has been established in the reactor. Both the regenerated catalyst slide valve and the spent catalyst slide valve should be closed.

Before removing the blind, the liquid inventory in the main fractionator should be brought to a minimum. The pressures in the main fractionator and reactor should also be minimized.

Once the transfer line flange is opened, the blind must be removed and the spacer ring and gaskets installed. Steam from the reactor and main fractionator will vent from the open flange and this generally makes this operation both difficult and uncomfortable. It is not uncommon for at least one gasket to be damaged during this step. It is a good practice to have an extra set of gaskets available at the site. Extraneous personnel should not be present during this operation as they will add to the difficulties.

Some refiners have installed a large gate valve between the transfer line flange and the main fractionator. This eliminates the need to remove hydrocarbon inventory from the main fractionator. Thus, the commissioning activities on the main fractionator are not interrupted. This results in a 12–24 hour decrease in total start-up time.[5]

During operation, coke builds up in the valve and this prevents closure during shut-downs. During turnarounds, the valve is removed and cleaned.

Other refiners have used systems involving hydraulic tools to open and close the transfer line flange. With these systems, the flange bolts can be removed while the flange is held in place by the hydraulic system. Once all of the bolts are removed, the flange is opened to allow the removal of the blind and the placing of the spacer and gaskets. The hydraulic tools are then used to close the blind and hold it in position while the bolts are replaced.

Transfer line coking is the primary problem normally encountered during normal operation. As discussed earlier, coking can occur due to heat loss from the transfer line. Coking can also be the result of poor feed vaporization in the riser.

If the coke deposits are localized, they are probably the result of localized heat loss from the line. Common problem areas are the transfer line flange, the transfer line vent and line support points. If the coke buildup is at the entrance to the main fractionator (after the transfer line flange), liquid entrainment from the tower may be the cause.

Poor feed vaporization generally results in coke deposits across the entire length of the transfer line. Poor vaporization can result from improper operation of the feed nozzles, especially operation at low atomizing steam rates. Operation at low reactor temperatures with high boiling point feeds can also result in poor feed vaporization.

Once coke has formed in the transfer line, it must generally be removed at the next turnaround. Transfer line coke is usually quite hard and adheres firmly to the transfer line walls.

Easily accessed localized deposits are generally removed by mechanical methods. Deposits at the transfer line flange are an example of this relatively easy-to-remove coke. Generalized deposits or deposits in difficult-to-reach sections of the line pose greater difficulties. At times, these deposits have forced partial or complete replacement of the transfer line.

These deposits can often be removed, however, by using high pressure water jets. A tank cleaning spray head attached to a flexible hose can often be used to clean coke from a transfer line. If a spray head with backward facing jets is available, the head will propel itself down the line.

Main Fractionator

Problems in the main fractionator generally occur in the bottom or in the top of the tower. In the bottom of the tower, coke formation can lead to problems with the bottoms circulating system. Other deposits can also collect in the slurry pumparound system leading to high pressure drops and reduced capacities.

Coke Formation. Coke formation in the bottom of the main fractionator results from a combination of high temperatures and long residence times. Under these conditions, the slurry oil undergoes thermal cracking reactions that eventually lead to the formation of coke. To minimize these problems, many main fractionators are designed with minimum liquid residence times in the bottom of the tower. In addition, the liquid in the bottom of the tower is often cooled by injecting cold slurry from the pumparound system.

The allowable operating temperatures in the bottom of the tower are a function of feed type and the properties of the slurry oil. Even at low temperatures, however, coke can still form if the slurry pumparound liquid is not well-distributed over the baffles. Poor liquid distribution will result in localized hot spots on the baffles. Coke will form on these hot spots. Eventually, these deposits will break free and fall into the bottom of the tower. They may also enter the slurry draw-off line and collect on the slurry pump strainers.

Coke formed on the baffles will often have a smooth side where the particle formed on the baffle surface. Large collections of this smooth-sided coke in the slurry system indicate a problem with liquid distribution in the desuperheating zone. This may be due to inadequate pumparound rates or to damaged tower internals. If possible, the slurry pumparound rate should be set at 1.2 to 1.5 times the feed rate. If this does not alleviate the problem, then the condition of the liquid distributors and the baffles should be investigated.

Coke or other deposits in the bottom of the tower and/or in the slurry pumparound system indicate thermal cracking of the bottoms product. This can be checked by running a distillation on a sample of the bottoms liquid.

The hydrocarbon products leaving the FCC reactor contain very little material boiling above 1100°F. Thus, a bottoms stream with a high fraction boiling over 1100°F is a strong indication of thermal cracking in the tower or in the slurry system. This thermal cracking results in the precipitation of large polynuclear aromatic compounds which combine with catalyst particles in the slurry to form tar-like deposits. Eventually, these deposits can react further to form coke.

If thermal cracking is indicated, the residence time of the slurry in the bottom of the tower and in the hot section of the pumparound system should be minimized. If possible, the temperature in the bottom of the tower should be reduced by increasing the amount of cold slurry injected into the tower bottom.

Various antifoulants have been used in slurry systems with mixed success. These chemicals are intended to prevent the precipitation of asphaltenes from the circulating streams. Finding the correct antifoulant is generally a trial and error process. This can be facilitated by close communication with the chemical vendor. The actual performance of the antifoulant can be judged by evaluating the decline in the heat transfer over time. The heat transfer decline should be determined for the base system (without chemical injection) prior to the beginning of the chemical test. This rate of decline should be measured on an exchanger that has just been cleaned.

Once the base rate of decline has been established, the test exchanger should be cleaned again and then returned to service. The test chemical should be injected as per the manufacturer's recommendations, and the decline in the heat transfer on the test exchanger should be determined again. The difference in the heat transfer decline between these two cases can be used to determine the value of the tested antifoulant.

Fouling in Main Fractionator Upper Sections. Fouling in the upper sections of the main fractionator and in the main fractionator overhead section is caused by salt deposits—usually ammonium chloride. These deposits form when water condenses in the top section of the tower.

Water condensation occurs when the temperature at the top of the column falls below the water dew point. This is generally due to the removal of a heavy naphtha side stream or a decrease in the cut point between the naphtha and light cycle oil products. Deposits formed in the overhead system can be removed by a well-designed water wash system. Deposits formed in the tower are, however, more difficult to remove.

Some symptoms of salt buildup are:

high pressure drop
early flooding
poor response to changes in reflux rate
loss of separation between gasoline and LCO

In severe cases, deposits may actually be found in product draw-off lines and pump suctions.

Salt deposits in the main fractionator are generally located in the naphtha section. In towers equipped with a heavy naphtha pumparound the deposits are generally located above the draw-off for this stream.

Salt deposits can be removed by on line water washing. On line water washing involves introducing water into the top of the main fractionator and allowing it to move down the column to the heavy naphtha section. This process will produce significant upsets in the column and should be undertaken with due care.

Wet Gas Compressor

The most common operational problem with the wet gas compressor is inadequate capacity. The most direct response to this problem is to raise the suction pressure. This increases compressor capacity, but also results in increased pressures in the main fractionator and the FCC reactor. If the FCC regenerator pressure is increased to balance the catalyst slide valve differentials, the air blower capacity will decrease.

Most FCCs were designed with a pressure balance intended to equalize the pressure drop across the spent and regenerated catalyst slide valves. This was done so that the wear rate on the valves would be equal.

While this approach is logical from the standpoint of mechanical design, it is not necessarily the best in operational terms. In addition, modern slide valve designs are extremely erosion resistant. Thus, balancing the wear between the two valves has little real benefit.

For units operating against gas and/or air blower restraints, the proper approach is to maximize reactor pressure and minimize regenerator pressure. These moves are generally subject to the constraints of minimum regenerated catalyst slide valve differential and maximum regenerator velocities.

This approach is often beneficial to units that are not operating against constraints as well, since this approach will minimize the power required by the air blower and the wet gas compressor. Units with flue gas power recovery are the one exception to this rule since the expander turbine will usually deliver the most power when operated at its design inlet pressure.

Many FCC operations impose an unnecessary load on the wet gas compressor by operating the sponge absorber to recover C_3s and C_4s. The purpose of the sponge absorber is to collect material lost from the primary absorber lean oil, not to recover LPG. Any LPG material collected in the sponge absorber is returned to the main fractionator where it is vaporized. Eventually, this "recycle gas" flows to the wet gas compressor.

The lean oil rate to the sponge absorber should be set to maintain the C_5 content of the absorber gas at or slightly below the maximum allowable level. C_3/C_4 recovery should be controlled by the recycle lean oil rate to the primary absorber.

Wet gas compressors are subject to fouling from deposits formed by polymerization of the olefins in the gas stream. These deposits usually form on the compressor rotor. This problem can be alleviated by a continuous compressor wash. The normal wash medium is light cat naphtha.

High Pressure Separator and Absorber/Stripper

As discussed earlier, this is actually a single separation system divided into two towers and a cooled feed stage. The flow of lean oil to the primary absorber is set to give the desired propylene recovery and the stripper reboiler operation is set based on the ethane in the C_3/C_4 LPG. The temperature in the high pressure separator is not normally controlled.

Overstripping and/or overabsorbing are common causes of upsets in this system. When the stripping rate is increased, the flow of vapor to the high pressure condenser increases. This increased vapor is either condensed, in which case it is returned to the stripper or it flows to the primary absorber via the high pressure separator.

If returned to the stripper, it will be revaporized and sent back to the high pressure condenser. If it flows to the absorber, it will be picked up by the lean oil and returned to the high pressure condenser. In either event, a recycle of material occurs. If the stripping rate is too high, this recycle stream can reach levels that overload the system.

Similarly, as the lean oil rate to the primary absorber is increased, the quantity of light material returned to the high pressure condenser increases. These increased light hydrocarbons can either be condensed and sent to the stripper or revaporized and returned to the absorber. They will be vaporized in the stripper and they will be reabsorbed in the absorber.

Overstripping or overabsorbing generally overloads the primary absorber first. The usual symptom is an inability to maintain the liquid level in the bottom of this tower. Once the problem occurs, the entire tower can quickly fill with liquid.

A similar overload can occur if the temperature in the high pressure separator is too high. In this case, the gas flowing to the absorber contains a high volume of easily condensed materials. These condense in the absorber and are returned to the high pressure separator where they are revaporized. As with overstripping or overabsorbing, the recycle builds up until it overloads the liquid handling capacity of the primary absorber.

Debutanizer

The primary operational problem in the debutanizer is a buildup of noncondensable gases in the overhead system. This is usually due to poor removal of the ethane in the stripper. When this occurs, gas from the debutanizer overhead drum must be vented to control pressure. This results in a loss of C_3 and C_4 products. There is little that can be done on the debutanizer to correct this problem. Instead, the operation of the stripper should be corrected.

Fouling on the process side of the debutanizer reboiler can occur due to polymerization of olefins and diolefins in the gasoline product. This can be minimized by avoiding excessively high temperatures in the reboiling medium.

Reboiler Control

The traditional approach to controlling reboilers heated by hot circulating oil streams is to bypass the hot oil to reduce the reboiler duty (Figure 8-18). This often did not provide adequate

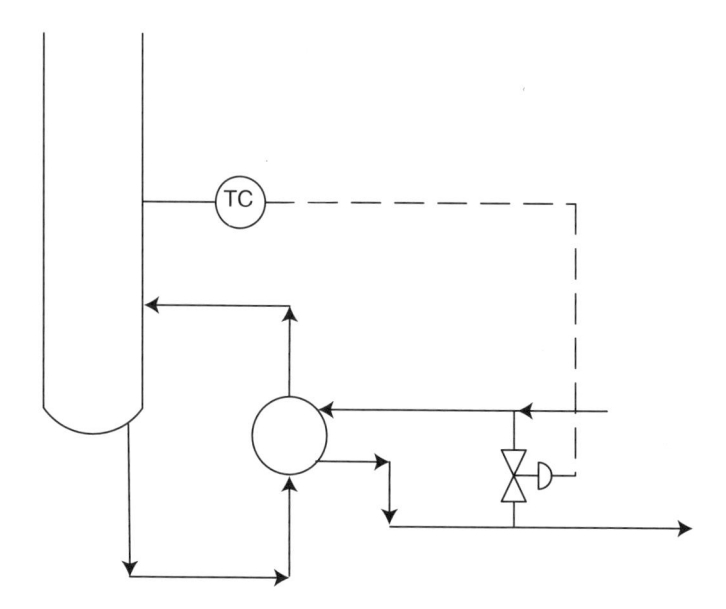

FIGURE 8-18. Conventional reboiler control.

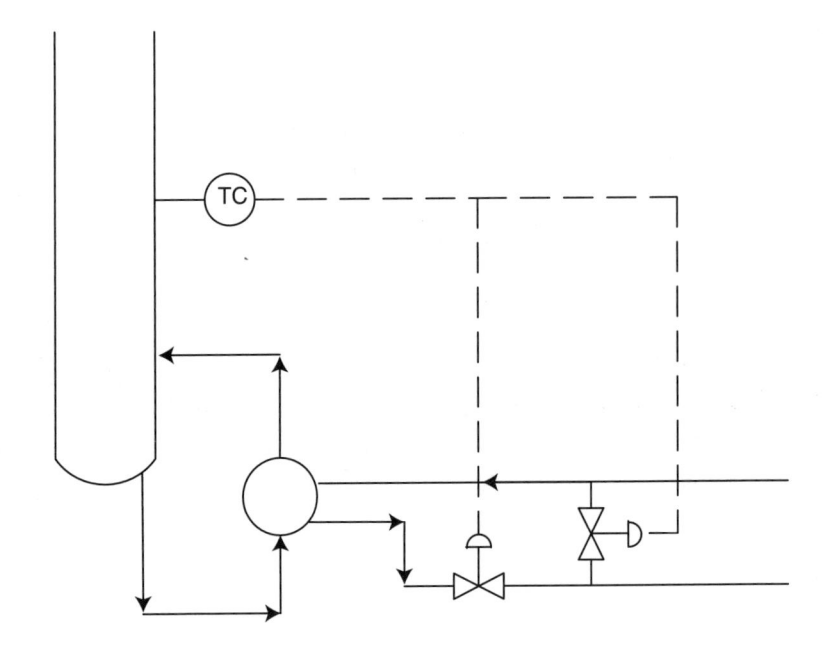

FIGURE 8-19. Improved reboiler control.

control since there was no positive control on the flow to the reboiler. Figure 8-19 shows an improved system that controls the flow through the reboiler and the flow through the bypass. This scheme will provide a more positive control of the reboiler.

References

[1] NPRA Question and Answer Session (1993).

[2] Strong, R.C., Majestic, V.K., and Wilhelm, M.S., "Basic Steps Lead to FCC Corrosion Control," *Oil and Gas Journal,* September 30 (1991).

[3] Sadeghbeigi, R., *Fluid Catalytic Cracking Handbook, Design Operation and Troubleshooting of FCC Facilities,* Gulf Publishing Co. (1995).

[4] Lieberman, N.P., *Oil and Gas Journal,* February 16 (1981), pp. 101–102.

[5] Andrews, P.G., Private Communication (1995).

9
Catalyst Technology, Selection, and Monitoring

Early Catalysts

Early FCC catalysts were naturally-occurring clays. Initially, catalysts were prepared by grinding, followed by screening to produce the desired particle size distribution. The introduction of spray-dried synthetic catalysts in 1948 greatly improved both the fluidization characteristics of FCC catalysts as well as their attrition resistance.

The introduction and commercialization of zeolite catalysts in the 1960s was a major milestone in the development of fluid catalytic cracking. Compared to synthetic amorphous catalyst, zeolitic catalyst offered improved activity retention as well as superior yield distribution. To take full advantage of this improved performance, the FCC process quickly moved from bed cracking to all riser cracking.

Today, zeolites are the primary catalytic ingredient in all FCC catalyst formulations. Zeolite technology has moved from the early X faujasites used in the 1960s to a wide range of variations on the more stable Y faujasite crystal.

In addition to zeolite, today's cracking catalyst contains several other components. These include active aluminas, binders, fillers, metals, traps, or materials intended to catalyze reactions other than cracking. Each of these components plays a role in the overall functioning of the catalyst. In making catalyst decisions, the FCC engineer must weigh the effects of each and estimate the performance of various combinations of ingredients.

Catalyst Technology

Catalysts contain different components each of which is intended to affect the catalytic and/or physical properties in the FCC. The primary components are zeolites, active alumina, binders, and fillers. Active aluminas, binders, and fillers are often referred to collectively as the catalyst matrix.

The relative percentages of each of these in the catalyst formulation will affect catalyst performance. In addition, each of these components may consist of one or more types and have been processed in one of several different ways.

Zeolite

Zeolites are the primary active ingredient in today's FCC catalyst. Without exception, zeolites are synthetically-produced Y faujasites. Proprietary zeolite names, used by catalyst vendors, refer to the manufacturing process used and not to the basic crystal type.

Zeolites are crystalline silica aluminates. The basic crystal structure is tetrahedral with a silicon or aluminum atom in the center of the tetrahedron and oxygen atoms at the corners. These basic building blocks combine at their oxygen atoms to form a cage structure that has well-defined pores of a specific diameter. The desirable catalytic properties of zeolites are the result of this regular crystal structure.

The silicon atoms in the zeolite crystal are electrically neutral. The aluminum atoms, however, carry a negative charge. This ionic site accounts for the catalytic activity of the zeolite.

The key property of the zeolite is the unit cell size. This is the length of the smallest repeating unit in the crystal structure. The aluminum-oxygen bonds are longer than the silicon-oxygen bonds. Thus, as the number of aluminum atoms in the crystal increases, the unit cell size increases. Since the aluminum atoms form the catalytically-active sites in the zeolite crystal, a higher unit cell size also indicates a higher activity.

In addition to the activity effects, increased unit cell size indicates the aluminum atoms, and thus the acid sites are closer together in the crystal. This means that bimolecular reactions such as hydrogen transfer are more likely to occur on a high unit cell size zeolite.

Increased hydrogen transfer reduces the olefin content of the gasoline produced and thus, both the octane number and the bromine number. This is most noticeable in the lighter gasoline fractions since these contain the most olefins. Increased hydrogen transfer activity also leads to a decrease in the olefinicity of the C_3 and C_4 LPG produced.

As explained in Chapter 2, hydrogen transfer reactions can also transfer the carbenium ion charge from a shorter molecule to a longer molecule. When this occurs, the shorter molecule is less likely to undergo subsequent cracking. For this reason, catalysts with increased hydrogen transfer activity will tend to produce higher yields of gasoline range material and lower yields of LPG.

A fourth effect of increased hydrogen transfer is an increase in the coke yield at any given conversion. This is due to the increased formation of aromatic molecules and their eventual condensation into multiring compounds. Thus, catalyst with a high hydrogen transfer activity will generally show higher delta coke and higher regenerator temperatures.

The zeolite unit cell size is controlled by the level of rare earth exchange on the zeolite. Zeolites are synthesized in a solution of sodium hydroxide and initially, the ionic sites on the aluminum atoms are occupied by positively charged sodium atoms. At this stage, the zeolite is referred to as sodium Y or NaY.

If left on the zeolite, these sodium ions would reduce the stability of the zeolite at typical regenerator conditions. This would have a negative effect on the long-term activity maintenance of the catalyst. The sodium ions are removed in an ion exchange with a solution containing other positive ions, usually hydrogen and/or rare earth.

Zeolite in which the sodium ions have been replaced with hydrogen are referred to as hydrogen Y or HY zeolites. Zeolites which have been exchanged with rare earth ions are referred to as rare earth Y or ReY zeolites.

The presence of rare earth on the ion exchange sites stabilizes the aluminum atoms in the crystal structure. This prevents the loss of aluminum atoms during subsequent heat treatment and later during exposure to high temperatures in the FCC regenerator. By holding the aluminum atoms in the zeolite structure, the rare earth limits the shrinkage of the unit cell. Thus, a zeolite with a high level of rare earth exchange will have a higher equilibrium unit cell size.

In addition to the equilibrium unit cell size, other important parameters in zeolite performance are the initial unit cell size and the type and degree of post crystallization treatment. In the initial crystallization step, the unit cell size is set by the operating parameters in the crystallizer (feed composition, seed composition and concentration, crystallizer temperature). This unit cell size is a function of the silica to alumina ratio of the sodium Y zeolite crystal.

Following the crystallization step, the unit cell size may be reduced through post crystallization heat treatment. This step is used to set the unit cell size as well as to fix the exchanged ions on the zeolite crystal. After crystallization and ion exchange, the zeolite is subjected to elevated temperatures in a calciner. In this step, some of the aluminum atoms in the zeolite leave the crystal and become free alumina. The degree of dealumination and thus, unit cell shrinkage, depends on the treatment temperature and the level of rare earth exchange.

It is important to note that while the aluminum atoms are removed from the zeolite crystal, they are not removed physically from the zeolite. Instead, these free or nonframework aluminum atoms form an alumina deposit in the zeolite surface. This deposit is catalytically active and has an effect on both catalyst activity and on yield selectivity.

Since the free alumina does not have the precise crystalline structure of the zeolite, it demonstrates poor yield selectivity. This results in increased yields of both coke and gas. For a hydrothermally treated zeolite with a specific unit cell size, a lower initial unit cell size will result in a lower percentage of free alumina in the final zeolite. This would be expected to produce improved yield selectivity in the finished catalyst.

Chemical treatments can be used to remove some of the free alumina from the zeolite. For this reason, catalysts containing chemically treated zeolites often exhibit superior yield selectivities.

Another advantage of chemically treated zeolites is improved stability. This phenomenon is not fully understood, but is believed to result from the removal of free alumina molecules from the zeolite surface. It is possible that this alumina can act to deactivate the zeolite through pore blockage. This same benefit can be at least partially obtained in hydrothermally processed zeolites by starting with a higher silica to alumina ratio and thus a lower initial unit cell size. Since less free alumina will be evolved during unit cell shrinkage, there will be less free alumina present to affect the zeolite.

Chemical treatments currently in use by catalyst manufacturers are considered proprietary and are not generally divulged to customers. Thus, the presence of a chemically-treated zeolite is difficult to determine.

Once a catalyst is placed in an operating unit, the zeolite will undergo further dealumination in the regenerator. As with hydrothermal dealumination in the manufacturing process, the aluminum removed from the crystal lattice does not leave the zeolite but collects as free alumina. For this reason, all equilibrium catalysts contain some nonframework alumina.

The quantity of zeolite in the catalyst can be estimated by measuring the relative zeolite crystallinity. This is done by X-ray diffraction. Since zeolite is a crystal, it will have a recognizable diffraction spectrum. The height of these spectral lines is a function of the zeolite concentration in the sample. The relative crystallinity is defined as the height of the zeolite spectrum in a catalyst sample expressed as a percent of the height of the spectrum from a sample of pure zeolite.

Matrix

The other constituent of the catalyst is the nonzeolitic portion or the matrix. This consists of binders, fillers, and nonzeolitic active ingredients.

Binders act to hold the particle together. The most frequently used binder is silica based. This binder is catalytically neutral. Other binders include natural or treated clays and various aluminum compounds. In addition to acting as a binder, most of these materials also have some catalytic activity.

Kaoline clay is the filler that is used to contain the active ingredients. The filler serves to provide the catalyst with its gross physical properties such as hardness, density, and size.

The nonzeolitic active ingredients, usually some form of active alumina, are used to modify the performance of the zeolite, and to provide additional activity and selectivity variations to the catalyst. The primary purpose of this active matrix is to crack the larger molecules present in the FCC feed. Most of the active sites in the zeolite are enclosed in the crystal supercage. The pores in the zeolite crystal have a uniform diameter and will not pass molecules larger than about eight angstrom units in diameter. Larger molecules must be precracked either on the external surface of the zeolite or by an active ingredient with larger pores.

Active alumina can provide these larger pores and thus improve the overall activity of the catalyst. This can be especially useful in FCCs that process heavy feeds such as resids.

In addition to providing improved cracking of the larger feed molecules, the active matrix can also act to suppress the formation of lighter products. This is done by providing additional hydrogen transfer activity to the catalyst.

Active matrix can also have a negative effect on yields in units processing resids or other feeds with high levels of nickel contamination. The nickel in the feed will deposit on the catalyst where it will catalyze dehydrogenation reactions. These produce increased yields of gas and increase the delta coke.

The dehydrogenation activity of the nickel on the catalyst is a function of both its age and its degree of dispersion on the catalyst particle. Alumina provides an excellent surface for metal dispersion. This is the reason that aluminas are used as a support medium for metal catalyst such as those used in catalytic reforming or hydrodesulfurization. Active alumina in an FCC catalyst can also act to disperse the nickel deposits and thus, increase their dehydrogenation activity.

Since both the zeolite itself as well as the kaoline clay filler contain alumina, the active matrix cannot be measured directly by chemical analysis. Instead, the usual procedure is to measure the matrix surface area. This is defined as the surface area contained in large diameter pores. The results of this measurement are normally reported as the ratio of zeolite surface area to matrix surface area or Z/M.

This measure is actually the ratio of the surface area present in small pores to that present in large pores. Thus, zeolite to matrix surface area ratio is based on the assumption that the zeolite pores are all small and the matrix pores are large.

Catalyst Technology, Selection, and Monitoring **269**

Unfortunately, this is an oversimplification. While the zeolite pore structure is fairly uniform, the matrix pore structure can contain both large and small diameter pores. In addition, matrix surface area can be either active or nonactive.

Thus, the Z/M ratio, while a useful parameter in evaluating catalysts, does not tell the full story with regard to matrix activity. In addition to the Z/M ratio, the weight percent alumina in the catalyst and the relative zeolite crystallinity should also be considered. Since active matrix components are high in alumina, a low Z/M ratio coupled with a high alumina content will generally indicate a high level of matrix activity. Even this, however, will only provide a general guide to the catalytic properties of the catalyst. Different types of active alumina have different yield and activity performance. In fact, many of the performance differences observed between similar catalysts from different vendors are the result of differences in the matrix. Thus, two catalysts with similar Z/M ratios and alumina contents can have widely different performance depending on the types of alumina used.

Catalyst Manufacture

Zeolite Synthesis

For most catalysts, the manufacturing process starts with zeolite synthesis. This is a batch crystallization process. In this step, silica and alumina are reacted in a caustic solution under a controlled temperature program. The most common sources of these materials are sodium silicate and sodium aluminate. Other sources can be used, however.

In addition to the raw materials, the charge to the crystallizers includes seed crystals which are grown in a separate process. The composition of these seeds is a critical parameter in the synthesis step.

The time required for crystallization will depend on the initial silica to alumina ratio and other operating parameters. Generally, higher silica to alumina ratios result in longer crystallization times.

The zeolite produced in the synthesis step contains sodium on the ion exchange sites. This sodium must be removed by ion exchange.

The ion exchange step involves washing the NaY crystals with solutions containing either ammonia or rare earth. Ammonia exchange is used to produce hydrogen Y zeolite while rare earths are used to produce ReY. There are several exchange steps, and the exchange solution can be different in each step. This permits easy control of the rare earth level on the zeolite.

Following the ion exchange step, the zeolite is dried and possibly calcined. In the calcination step, the time and temperature are controlled to achieve the desired unit cell size. The calcined zeolite may then be chemically treated to remove some of the free alumina evolved during calcination or to produce other modifications of its catalytic properties.

Alumina Synthesis

Active matrix alumina can either be purchased or can be synthesized. Various types of alumina are used by various catalyst vendors. Each has its own specific properties.

Catalyst Formulation

The zeolite, active alumina, and kaoline clay are first mixed with binder and this slurry is fed to a spray dryer. In the spray dryer, the slurry is forced through nozzles and discharged as a high velocity thin film into the dryer chamber. Hot air is also fed to the dryer chamber.

The film produced by the nozzles is unstable, and on contact with the relatively stationary air in the dryer, it breaks up into small droplets. These droplets then harden into catalyst particles.

The composition of the slurry fed to the dryer determines most of the physical properties of the catalyst. If the slurry contains a high percentage of zeolite and/or alumina, the catalyst will tend to be lower in density and less attrition resistant. This can be compensated for by increasing the percentage of the binder in the mix. Too much binder, however, can have a detrimental effect on catalyst performance since the binder can obstruct the catalyst pores.

Proper operation of the spray dryer is essential to achieving the desired physical properties of the catalyst. Spray dryer operation affects the particle size distribution, attrition index, and even the shape of the catalyst particles. Ideally, the particles formed are spherical. In some cases, however, the particles formed will contain irregularities. These uneven surfaces will tend to break off in the unit leading to higher attrition rates.

Catalyst vendors today make extensive use of statistical process control to monitor the performance of the spray dryer as well as other critical steps in the manufacturing process. This practice has resulted in increased uniformity of fresh catalyst properties.

Finishing Steps

The catalyst from the spray dryer contains sodium and other undesirable compounds. These are removed by washing. In some cases, additional ion exchange with either ammonia or rare earth is performed in this step.

After washing, the catalyst is dried. In some cases, the catalyst may also be calcined. This final calcining step provides two benefits. First, this step increases the hardness of the catalyst particle. Secondly, the final calcining step reduces the water content of the catalyst and decreases the catalyst's tendency to absorb moisture from the atmosphere.

These effects reduce the tendency of the catalyst to break up into small particles when it is first loaded into an operating FCC. Uncalcined fresh catalyst is soft relative to equilibrium catalyst. Thus, a high fraction of the catalyst lost from the unit is often fresh catalyst that has not had time to harden in the unit. In addition, fresh catalyst normally contains between 14 and 15% moisture by weight. This water is lost when the catalyst is fed to the unit and this rapid drying can cause some attrition of the catalyst.

Engelhard Process

The manufacturing process described above is typical of the catalyst produced by most vendors. Engelhard, however, produces most of its catalyst by in situ growth of zeolite in clay spheres.

In the Engelhard process, kaolin clay is first spray-dried into microspherical particles. These microspheres are then calcined at very high temperatures (c.1800°F) to produce very hard particles. Zeolite crystals are then grown in the pores of the microspheres.

As with the zeolite produced in more conventional crystallization, the zeolite grown in the microspheres is a NaY. The sodium is removed by ion exchange of the catalyst particle.

Catalysts produced by this process always contain an active matrix. They are also very hard and show excellent resistance to attrition.

Engelhard also produces catalysts by the more conventional process used by other catalyst manufacturers.

Catalyst Testing and Evaluation

Modern catalyst technology is very complex and there are a wide variety of catalysts on the market. Catalyst formulations and performance can be adjusted by changing the amount and type of zeolite used, as well as by varying the composition of the nonzeolite matrix. Further modifications can be made based on the types of treatment used on both the zeolite and the finished catalyst.

Given this flexibility in manufacturing, it is not surprising that catalyst suppliers have introduced such a large number of different catalysts. At times, the number of catalyst types available has actually exceeded the number of operating FCCs in the world.

Catalyst manufacturers are understandably reluctant to discuss the details of their manufacturing process or catalyst formulations. This, coupled with the wide variety of catalysts available, makes rational selection decisions difficult for even knowledgeable engineers. From a refinery operations standpoint, however, the proper choice of catalyst is crucial to optimum FCC performance. For this reason, an objective and scientific process for evaluating catalysts is essential.

Often, however, catalysts are evaluated based entirely on the vendors' estimates. The yields predicted by each vendor are placed into an economic evaluation model along with the catalyst costs. The results of this comparison are then used to select the "best" catalyst.

Since the suppliers' yield estimates generally show an improvement over the current operation, this procedure, in effect, makes the FCC the test unit to verify the claims of the selected supplier. This is a very accurate test procedure since the test results are the actual unit performance. The potential costs are, however, high.

This potential for losses is compounded by the fact that it will often be several weeks before the results of a poor catalyst choice will be recognized. Even then, the losses will not stop, since it will take some time to select and obtain a replacement catalyst, and then, additional time will be required to replace the poor-performing catalyst.

These problems can be avoided by independent catalyst testing and evaluation. This practice has gained wide acceptance throughout the refining industry, especially among larger companies with more than one refinery.

Test Methods

Catalysts can be tested using bench scale or pilot plant sized equipment. Bench scale tests are conducted using fixed bed test equipment derived from the standard microactivity test unit. These units use a small (approximately 1 gram) catalyst sample in an isothermal fixed bed. Feed oil is forced through the bed at a controlled rate. Liquid and gaseous products are analyzed by gas chromatograph, and the coke yield is determined by carbon analysis of the used catalyst.

In these tests, conversion is varied by changing the catalyst in the fixed bed or by changing the oil injection time. Most test procedures change the quantity of catalyst. Tests of this type have the advantage of being fast and relatively inexpensive. Unfortunately, the results from bench scale tests have limited value.

The conditions in the isothermal fixed bed do not resemble the conditions found in an operating riser cracker. In the fixed bed test, the feed/product mixture moves through a bed of freshly regenerated catalyst. Thus, the mixture continuously sees clean catalyst. In a riser, the feed/product mixture and the catalyst flow up through the riser together. As the feed reacts, it sees progressively more carbon on the catalyst.

In addition, due to the endothermic nature of the cracking reactions and the low heat transfer rate across the walls of the riser, the reaction mixture cools significantly as the reaction proceeds. Typically, the temperature in the riser will fall between 50–70°F from the feed injection point to the riser outlet. In a bench scale unit, the catalyst bed is held at a constant temperature throughout the test.

These differences affect both reaction rate and the yield selectivities. Thus, results from the bench scale test may not truly reflect the performance of the catalyst in a commercial FCC.

Pilot plant testing reduces or eliminates many of the problems associated with bench scale testing. In an FCC pilot plant, the feed is contacted with hot catalyst in a dynamic reaction system. This reaction system may be a fluid bed, fast fluid bed, or dilute phase riser. Of these, the true riser pilot plant is the closest simulation of a modern catalytic cracking unit.

In a riser pilot plant, the feed and catalyst move up the riser at essentially the same velocity. This is a close approximation of the reaction system in an operating cat cracker. In addition, through the proper use of insulation and heat control, a riser pilot plant can be operated in an adiabatic mode, thus simulating the temperature effects seen in a commercial unit.

Pilot plant testing also produces usable quantities of liquid products. As an example, the gasoline produced in a pilot plant test has sufficient volume to be tested in an octane engine. This is an important consideration when product quality is a major concern.

Pilot plant testing is, however, considerably more expensive than bench scale testing. Thus catalysts should be prescreened before being submitted for testing. This step will eliminate any obviously inappropriate catalyst from testing.

Evaluation Procedure

Due to the importance of catalyst selection to the proper operation of the FCC, catalyst selection should be based on a well-defined program that addresses frequency and evaluation procedures.

Frequency. A catalyst change should be considered whenever there has been a significant change in unit operations. This may be due to a change in operating objectives or it may be due to a significant change in the feed type to the FCC. These changes are generally initiated by the refinery planning department or by external factors related to markets and crude availability.

A catalyst change should also be considered following major unit modifications. In this case, it will first be necessary to operate the revamped unit on the prerevamp catalyst in order to establish a reasonable base case for the catalyst evaluation.

Refineries that operate with infrequent changes in feed and/or operating objectives are frequently tempted to remain on the same catalyst for extended periods. These refineries may benefit from more frequent catalyst switches.

Catalyst manufacturers are continuously developing new catalyst formulations as well as improved catalyst components. Each of these changes produces a catalyst with small but significant changes in performance. Thus, over time, better catalysts become available. To fully realize these possibilities, refiners should at least consider a catalyst change once every two years.

Evaluation

The evaluation process begins by requesting catalyst proposals from the major vendors. The vendors should be supplied with information on the current operation of the FCC and on the required performance criteria for the new catalyst. Based on these data, the vendors should prepare a catalyst proposal that includes the following information:

Fresh Catalyst Data
Type
Physical Properties
  ABD
  surface area
  pore volume
  unit cell size
  Z/M surface area
  particle size distribution
Chemical Properties
  alumina wt.%
  rare earth wt.%
  sodium wt.%
  iron
  sulfates
Catalytic Properties
  activity

Equilibrium Catalyst Properties
  activity
  ABD
  surface area
  Z/M surface area
  unit cell size
  metals
  specific coke
  specific gas
Performance Data
  yields (including complete weight balance)
  operating conditions
  product properties (as needed)

The next step is to screen the proposals to eliminate any obviously unsuitable catalysts. This is especially important if more than one catalyst is submitted by each vendor. The purpose of the screening step is to reduce the number of catalysts that will be tested and thus, the overall cost of the evaluation program.

Catalyst screening can be done based on the information supplied by the catalyst vendor and personal knowledge of catalyst performance. At times, consultation with the vendor regarding the reasons he proposed specific catalysts will also be of value. If a large number of samples must be screened, bench scale (MAT) testing can be a valuable tool.

The length of the short list will depend on the amount of time and money allocated for testing. Generally, however, it should contain at least one catalyst from each vendor.

Testing. The catalysts on the short list are tested to establish their relative performance against the current or base catalyst. This testing is best done in a riser pilot plant. MAT testing is less accurate but can be used if there are cost or availability restrictions to pilot plant testing.

Before testing, the catalyst must be deactivated. This is normally done by steaming. Steaming conditions are set so that the steamed sample of the current catalyst has similar properties to the unit equilibrium catalyst. These conditions are then used to deactivate the other catalysts.

The steamed catalysts are then tested using the actual unit feed. Tests with standard feeds are less valuable since the relative performance of the catalyst will change with variations in feed properties.

Sufficient data should be collected to show the relative performance of each catalyst. This often requires data from multiple conversion levels.

Depending on specific unit operations, other tests may also be required. Chief among these is testing for metals tolerance. This test is performed by impregnating the fresh catalyst with metals. Catalysts impregnated with several levels of metals are then tested for performance and these data are used to establish relative metals' tolerances.

The traditional method added the metals to the catalyst by impregnation with metal naphthenates followed by steaming. This results in all of the metals on the catalyst being the same age.

In a commercial unit, however, the metals on the catalyst are built up over successive passes through the riser. Thus, the metals on the catalyst have an age distribution. Research has shown that the effects of metals on the catalyst change as the metals age. Thus, single-step impregnation does not truly replicate the effects of metals on equilibrium catalyst.

This problem has been addressed by the development of cyclic deactivation units. These units are designed to impregnate the catalyst with metals in a series of steps. The catalyst is first exposed to a metals-loaded oil. Conditions in this step are similar to those found in a commercial reactor. The catalyst is then exposed to an oxidation step at conditions similar to those found in the regenerator. This cycle is repeated until the desired level of metals on the catalyst is reached.

This process provides a much better measure of the effects of metals on equilibrium catalyst. It does, however, require specially built equipment. In addition, the time required for cyclic impregnation is much longer than that required for batch impregnation. Cyclic impregnation units are generally small-scale and can only prepare samples for bench scale tests. Despite these drawbacks, cyclic impregnation is generally accepted as the best procedure for evaluating catalyst metals tolerance.

Due to the small sample sizes available from cyclic impregnation, performance tests are generally done using bench scale equipment.

Normalization. Data from the pilot plant and bench scale testing units must be normalized to actual unit operations. This step is necessary because the operating conditions in the test units differ significantly from those found in a commercial FCC.

In a commercial FCC, the catalyst circulation, conversion, and yields are set by the unit heat balance and this is driven by the coke yield. Thus, differences in coke selectivity between catalysts will result in changes in catalyst circulation, conversion, and product yields.

In a pilot plant, however, the catalyst circulation is controlled externally and the heat balance is set by the use of heating elements. Thus, the conversion and yields for each catalyst are not as dependent on its coke selectivity.

These differences are normalized using a heat balance model that has been tuned to the operation of the commercial unit. Ideally, this tuning should be done using test run data on the test feed and base catalyst. The normalized yields produced in this step are then used along with the catalyst price to determine which catalyst will give the best overall return.

Follow-up. Once a new catalyst has been selected and added to the unit, its performance should be closely monitored. Deviations from expected performance should be identified and discussed with the catalyst supplier. These deviations can often be explained by differences between the test conditions and unit operations. In many cases, they can be corrected by small changes in catalyst formulation.

Catalyst Monitoring

Routine monitoring of the catalyst in the FCC is necessary to insure that catalyst performance and, consequently, unit operations are as intended. In addition, routine monitoring of catalyst properties can provide early warnings of developing problems in the FCC. If recognized and acted upon, these early warning signs can minimize or prevent serious equipment damage or operational upsets.

Catalyst properties fall into three main categories: performance data, physical properties, and chemical properties.

Performance Data

Typical catalyst performance data are activity, gas factor, carbon factor, and dynamic activity. These data are measured in the microactivity test (MAT) unit. In the MAT, the catalyst is tested using a standard gas oil feed at standard conditions. While an ASTM procedure for this test does exist, it is not widely used. Instead, each testing laboratory has developed its own standard feed and conditions. MAT testing is done on clean catalyst. Any carbon present on the sample is burned off before the test.

Activity. The MAT activity (often called just MAT) is the conversion based on a 430°F TBP cut point gasoline product. The activity is a measure of the catalyst's ability to crack heavy oils to lighter products. A decrease in activity indicates a loss in the catalyst's ability to promote cracking reactions. This may indicate an increase in the level of metals contamination on the catalyst due to a change in feed quality and/or catalyst make up rate. It may also indicate a mechanical problem in the regenerator such as a damaged air or catalyst distributor or excessive steam in the regenerator bed.

Gas Factor. The gas factor is a relative measure of a yield of C_2 and lighter gas produced in the MAT. The reported factor is generally expressed as a fraction of the gas yield that would be obtained from a base catalyst at the same conversion. An increase in gas factor may indicate an increase in catalyst metals, or it may indicate a change in the catalyst formulation which increases the dehydrogenation activity of the metals.

Coke Factor. The coke factor is a measure of the coke yield produced by the catalyst as a fraction of the coke yield that would be produced by the base catalyst at the same conversion. Changes in the coke factor may be due to changes in the catalyst physical or chemical properties or to an increase in catalyst metals.

Dynamic Activity. The dynamic activity is a measure of the coke yield of the catalyst relative to the conversion. It is calculated by the following formula:

$$DA = \frac{\text{Second Order Activity}}{\text{Coke Yield, wt.\%}}$$

where:

$$\text{Second Order Activity} = \frac{\text{Activity}}{(100 - \text{Activity})}$$

and the coke yield is from the MAT. The dynamic activity is a measure of the catalyst's ability to convert feed at a constant coke yield. Since the coke yield in a commercial unit is a function of the unit heat balance, a decrease in the dynamic activity will generally result in a loss of conversion on the FCC.

Physical Properties

Physical properties include surface area (SA), pore volume (PV), average bulk density (ABD), particle size distribution (PSD), and average particle size (APS).

Surface Area. The surface area is measured in m^2/gm. For a given catalyst type it is an indirect measurement of the catalyst activity. The surface area measurement can be broken down further into zeolite and matrix surface area. As discussed earlier, this measurement is actually based on the surface area found in small pores (presumed to be zeolite) and large pores (presumed to be matrix). A loss of surface area indicates a loss of catalyst activity. This may indicate increased hydrothermal or thermal deactivation in the unit due to a steam leak in the regenerator. Other possible causes include increased catalyst metals or internal damage in the unit.

Pore Volume. The pore volume is measured as cc/gm. Generally, catalysts with higher pore volumes have higher levels of active matrix components. A decrease in pore volume may indicate thermal deactivation (hydrothermal deactivation has little effect on pore volume). This may be due to changes in unit operation or may indicate damage to the regenerator air or catalyst distributor.

Average Bulk Density. The ABD is measured in g/cc and is a function of the catalyst composition and of other physical properties such as pore volume and particle size distribution. A change in ABD generally indicates a change in these properties. Catalyst with high bulk densities are often difficult to fluidize. Excessively low ABDs can lead to increased catalyst losses.

Particle Size Distribution. The PSD can be measured by screening or by light scattering. Most measurements today are by light scattering. The PSD is a function of cyclone performance, catalyst properties, catalyst make-up rate, and unit operating conditions. Changes in PSD often indicate emerging problems in the unit. A decrease in the percentage of smaller particles (fines) generally indicates deteriorating cyclone performance. An increase in fines may indicate increased catalyst attrition due to internal damage.

Average Particle Size. The APS is the weight average particle size of the catalyst. In practice, however, the number reported is actually the median particle size.

Chemical Properties

Chemical properties include the alumina content, rare earth content, carbon level, and metals (Ni, V, Na, Fe, Cu) content.

Alumina Content. The alumina content is determined by the catalyst formulation. Changes in alumina content can be used to track catalyst change-out. Unexpected changes indicate a change in catalyst formulation.

Rare Earth Content. Rare earth content is also determined by the catalyst formulation. It can also be used to track catalyst change-out or as an indicator of catalyst formulation changes.

Carbon Level. The carbon level on the equilibrium catalyst is an indicator of regenerator effectiveness. Units operating in complete CO combustion generally have carbon levels of 0.05 wt.% or less on regenerated catalyst. Units operating in partial combustion generally have levels of 0.10% or higher. The carbon level on the regenerated catalyst has a direct impact on unit performance. Increased carbon levels reduce the effective catalyst activity. This effect can be estimated by the following formula:

$$MATC = MAT - 5.505 C_{rc} - 5.081 C_{rc}^2$$

where:

$MATC$ = effective catalyst activity (with carbon)
MAT = measured catalyst activity (carbon free)
C_{rc} = carbon on regenerated catalyst, wt.%

Metals. Metals found on equilibrium catalyst include sodium, nickel, vanadium, iron, and copper. The sodium found on the equilibrium catalyst comes from both the fresh catalyst and from sodium in the feed. Typically, the sodium levels in the fresh catalyst are a function of zeolite content.

Vanadium is deposited on the catalyst from the feed oils, especially residual feeds. This metal reacts with the zeolite in the catalyst and destroys the crystal structure. This leads to an irreversible loss of catalyst activity.

Nickel promotes dehydrogenation and when deposited on FCC catalyst leads to increased yields of hydrogen and coke. Iron and copper also promote dehydrogenation reactions.

The metals deposited on the catalyst can be estimated based on the feed rate, feed metals content, and the catalyst make-up rate. All of the metals in the feed are assumed to deposit on the catalyst in the riser. Thus, the metals on the catalyst are given by:

$$M_c = \frac{F \times M_F}{C}$$

where:

M_c = metals on catalyst, ppmw
F = feed rate, lb/hr
M_F = metals in feed, ppmw
C = catalyst addition rate, lb/hr

Monitoring

Catalyst monitoring is an essential aspect of FCC operations. Unfortunately, it is an aspect that is often neglected. The causes of this neglect are due to the nature of catalyst data and the demands placed on responsible engineers.

Catalysts are generally tested for the refinery by the catalyst vendor. Catalyst samples are collected on a regular schedule and sent to the supplier for analysis. The data from these tests are returned to the refinery in the form of an equilibrium catalyst data sheet. Typical data found on these sheets are shown in Figure 9-1.

Properly used and analyzed, the data on this sheet can provide considerable insight into the status of the FCC. It can also be used to detect emerging trends that may indicate the early stages of a serious operational or mechanical problem.

Catalyst Technology, Selection, and Monitoring 283

ACTIVITY	Gas Factor	Carbon Factor	S.A M2/GM	P.V CC/GM	ABD GM/CC	0-20 WT.%	0-40 WT.%	0-80 WT.%	APS	Al2O3 WT.%	Na WT.%	C WT.%	V ppmw	Ni ppmw	Cu ppmw	Sb ppmw
62	1.3	3.6	132	.33	.88	2	7	53	77	38.3	.30	.04	1543	842	21	-
61	1.6	4.2	131	.32	.90	0	6	52	76	38.1	.30	.05	1555	849	22	-
63	1.2	3.9	132	.32	.89	1	4	53	77	38.0	.32	.06	1566	855	23	-
66	1.1	3.1	137	.31	.90	0	5	56	78	38.2	.30	.05	1546	839	21	-
63	1.3	3.3	133	.32	.89	0	7	54	73	38.0	.30	.05	1561	852	22	-
62	1.4	3.8	133	.33	.89	1	6	57	76	38.2	.32	.06	1559	845	20	-
65	1.2	3.2	135	.31	.90	0	4	56	74	38.5	.33	.04	1563	850	23	-
64	1.3	3.3	134	.32	.90	0	5	56	74	38.2	.34	.05	1569	855	23	-
63	1.4	3.0	133	.33	.88	1	7	55	75	37.9	.34	.07	1554	845	20	-

FIGURE 9-1. Equilibrium catalyst data.

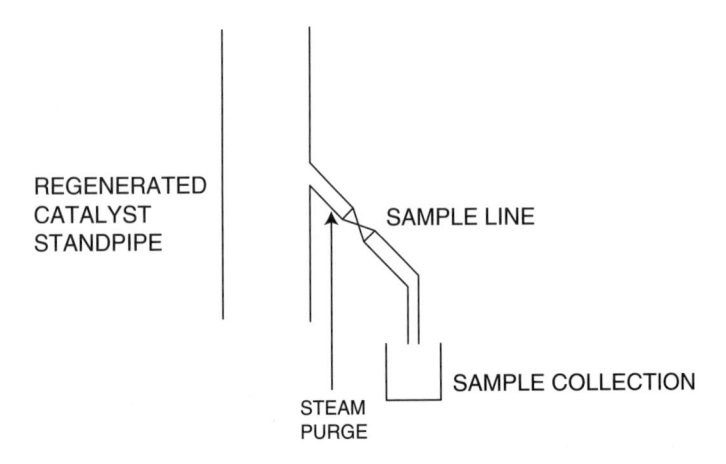

FIGURE 9-2. Catalyst sample point.

This analysis is, however, difficult. The data are subject to considerable random variation. This can be due to variations in sampling technique, handling of the sample once it has been collected, or from the accuracy range of the testing procedures used.

Catalyst sampling is a frequent cause of variation. A typical sample system is shown in Figure 9-2. In taking a sample, the operator will first purge the sample line with steam and will then open the sample valve. The catalyst is collected in a bucket or ladle. This sample is allowed to cool before being packaged and sent to the vendor for analysis.

Sampling procedures are usually not defined rigorously. Details are often left to individual operator preference. Thus, each step is subject to variation based on the individual operator.

Once the sample is collected it must be handled further, both in the refinery and in the vendor's labs. Since equilibrium catalyst is a mixture of particles of different sizes and ages, this handling can cause stratification. Only a small fraction of the total sample is used for each test. Consequently, if the catalyst is not well-mixed, these small samples may not reflect the properties of the complete sample.

Finally, each analytical procedure has its own range of accuracy. This will add further variation to the final results.

Once the data are returned to the refinery, the FCC engineer must decide if any trends are apparent. This is normally done by either visually scanning the information or by plotting it on a trend line. Due to the random variation in the results, however, these procedures will often fail to identify a trend until it is well established.

Statistical techniques can be used to separate emerging trends from the random variations in equilibrium catalyst data. This approach greatly increases the value of these data and thus, greatly improves catalyst monitoring.

Statistical monitoring of data is commonly used in manufacturing processes to identify deviations in product properties or other quality parameters that are beyond the normal random variation of the manufacturing process. These techniques are generally referred to as statistical process control. The same procedures can be used to identify trends in equilibrium catalyst data.

Most statistical process control systems use control charts constructed by plotting the value of the property in question against its mean value. The mean is calculated for a period when the process is known to be in control. That is, there are no unexpected causes of variation. Normally, at least 30 data points are used to calculate this mean. The standard deviation for this data sample is also calculated. The standard deviation is defined as:

$$\frac{\left(\sum_{i=1}^{n}(x_i - \bar{x})^2\right)^{0.5}}{n-1}$$

where:

n is the number of samples.

The calculated mean is plotted as a horizontal line on the chart. Control limits are plotted above and below the mean. Normally, the control limits are set at approximately 3.0 standard deviations.

Actual data are plotted on the control chart and the position of these points relative to the mean and control limits determines if nonrandom variation from the mean exists.

Control charts can be used to evaluate equilibrium catalyst data. The interpretation of these charts is somewhat subjective however.

Another statistical procedure which is somewhat easier to use and which gives clear indications of an out-of-control-limits situation is the CUSUM chart.[1] The first step in constructing a CUSUM chart is to calculate the mean and standard deviation of a family of samples. As with the more traditional control charts, these samples should be from a period of normal in-control operation.

The deviation of any future data point from the mean is calculated by:

$$D_i = x_i - \bar{x}$$

These deviations are then summed for each successive data point to give the cumulative sum of deviations or CUSUM.

$$CUSUM = \sum_{i=1}^{n} D_i$$

This value is then plotted. The slope of the CUSUM line indicates the current state of the process. A slope of zero indicates no significant deviation. A positive slope indicates a positive deviation from the normal average and a negative slope indicates a negative deviation. This allows for easy identification of trends.

Both graphical and numerical techniques are available to identify significant deviations.[1] These techniques can be used to confirm that an apparent trend is in fact due to a nonrandom cause.

Example—Catalyst Surface Area

The use of CUSUM charts can best be illustrated by an example. The catalyst surface area is an important parameter that can be used to track catalyst activity. Surface area measurements are relatively easy to perform and do not require the time or effort needed for a microactivity test. For a given type of catalyst, the activity will generally correlate very well with the surface area.

Table 9-1 contains typical surface area values. A visual scan of these data shows no obvious trend. A time plot (Figure 9-3) shows considerable variation with a possible rising trend toward the end of the plot. Faced with these data, the FCC engineer must decide if there has, in fact, been a change in surface area and thus in activity.

Figure 9-4 shows a CUSUM plot of the same data. For this plot, the first 14 points were used to calculate the mean and standard deviation. The CUSUM plot shows a definite trend beginning at about sample number 15. The positive slope of the line indicates that the surface area may indeed be rising. The procedure outlined by Butler gives a clear action signal on point 18. This indicates that the apparent trend is real and is the result of

Catalyst Technology, Selection, and Monitoring **287**

TABLE 9-1. Catalyst surface area.

Sample No.	Surface Area
1	133
2	132
3	134
4	135
5	131
6	131
7	133
8	130
9	131
10	131
11	130
12	132
13	133
14	132
15	135
16	133
17	134
18	136

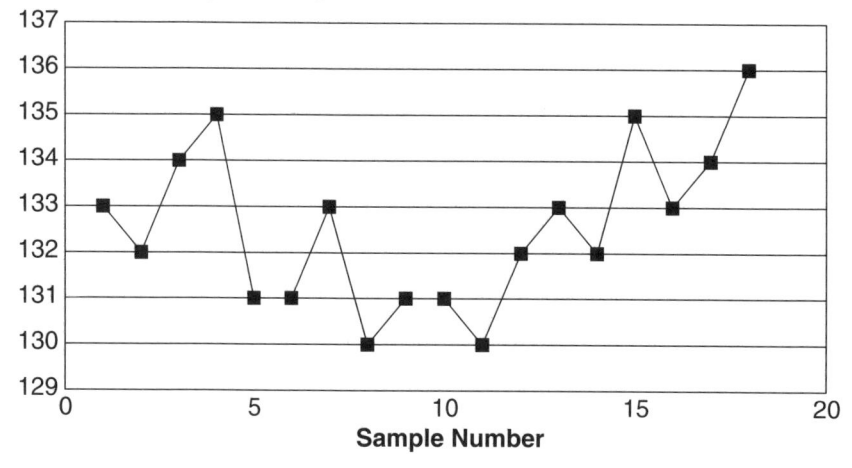

FIGURE 9-3. Catalyst surface time plot.

CUSUM

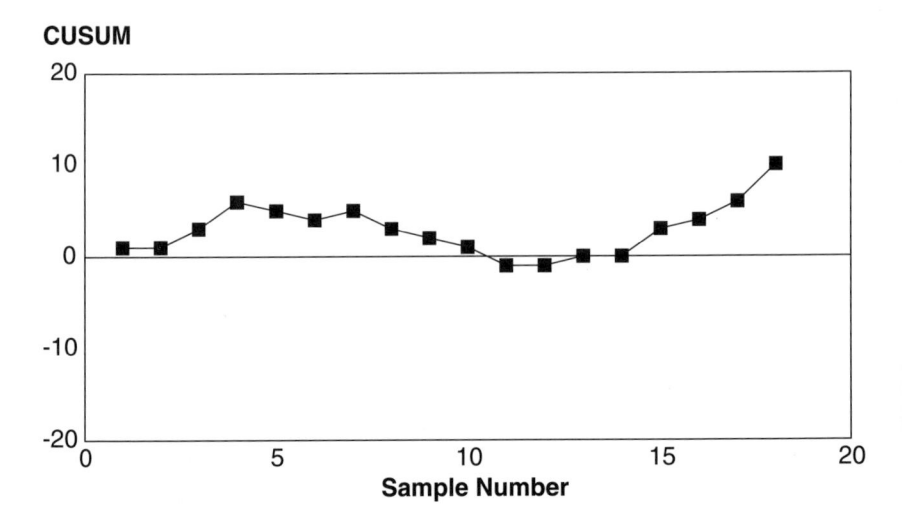

FIGURE 9-4. Catalyst surface area CUSUM plot.

some nonrandom change in the unit or catalyst. Since catalyst activity correlates well with surface area, this indicates that catalyst activity is also rising.

Most FCCs attempt to operate at an optimum catalyst activity. This is set by balancing the positive effects of higher activity, such as improved yield selectivity, against such negative effects as higher delta coke and increased catalyst costs.

The increased surface area is an indication that the catalyst activity may be deviating from the norm. To avoid possible negative consequences, the cause of this deviation should be identified and corrected.

This example illustrates the ability of CUSUM charts to detect small changes in catalyst properties. In addition, it illustrates another advantage. Compared to the time plot of the catalyst surface area, the CUSUM chart presents a much smoother curve. This makes visual identification of changes much easier. In many cases, this visual identification of an emerging trend is so strong that a full analysis of the CUSUM chart is not required.

Catalyst Additives

In addition to the base FCC catalyst, catalyst additives are available that can be used to modify the operation of the units or to adjust the yields or product properties.

ZSM-5

ZSM-5 is a shape-selective zeolite developed by Mobil. The crystal structure of ZSM-5 results in a pore structure that is significantly different from the Y faujasites used in cracking catalyst. ZSM-5 preferentially cracks long, straight chain hydrocarbons, either paraffins or olefins. The products of this selective cracking are shorter chain olefins.

Since long, straight chain hydrocarbons have low octane numbers, ZSM-5 can be used to crack these molecules into LPG and into shorter, higher octane gasoline molecules. Thus, ZSM-5 containing additives can be used to increase the octane number of the FCC gasoline product and to increase the yield of C_3 and C_4 olefins. These can also be converted into high octane gasoline components through further processing.

Since some gasoline is converted to LPG, the yield of gasoline decreases. In addition, since ZSM-5 cracks only paraffins and olefins, the remaining gasoline will be more aromatic.

There are two types of ZSM-5 currently available. The first type has a higher cracking activity and thus produces higher yields of light olefins and larger losses in gasoline yield. The second type has less cracking activity. This form tends to isomerize straight chain molecules to form branched hydrocarbons. Thus, the gasoline octane is increased with less loss in yield.

Combustion Promoter

Combustion promoters were also developed by Mobil. These additives contain platinum and/or other metals that promote the combustion of carbon monoxide to carbon dioxide. Combustion promoters are widely used in units practicing complete CO combustion. In these units, they confine the oxidation of CO to the bed and prevent high temperature increases in the regenerator dilute phase or cyclones. They are also used to control afterburn in units operating in partial CO combustion.

Combustion promoters are generally manufactured by dispersing the oxidation catalyst over an alumina support. Alumina is the preferred support since it provides the best dispersion of the metals over the particle.

SO_x Reduction

SO_x reduction additives are used to reduce the sulfur oxides in the flue gas. These additives work by adsorbing sulfur trioxide and reacting it with a metal oxide to form a metal sulfate. The metal sulfate is carried with the catalyst to the reactor where it is reduced to hydrogen sulfide. The net result is a reduction of sulfur oxides in the regenerator and an increase in the yield of H_2S.

SO_x reduction additives work best in the presence of significant excess oxygen in the regenerator flue gas. Thus, they are most effective in units operating in complete CO combustion. The additives are generally less effective in partial burn units but will still reduce the flue gas sulfur content somewhat.

To remain effective, the additive must regenerate well in the reactor. Some early additives were very effective at collecting sulfur in the regenerator, but did not release this sulfur in the reactor. These additives showed rapid loss of collection efficiency as they became saturated with adsorbed sulfur compounds. This rapid loss of effectiveness could only be countered by frequent additions of fresh material.

While many metal oxides can be used to adsorb sulfur in the regenerator, additives based on magnesium are by far the most effective. These additives give both excellent adsorbtion of sulfur oxides as well as excellent regeneration in the riser. Additives based on other metals generally have poorer regeneration and thus require higher make-up rates. These additives may, however, be cost effective for units with high overall catalyst make-up rates.

Bottoms Conversion Additives

Bottoms conversion additives are essentially catalyst particles that contain active matrix components, but no zeolite. These additives can be used to increase the yield of light cycle oil or gasoline at the expense of fractionator bottoms.

Metal Passivators

Strictly speaking, these are not catalyst additives but are instead liquids that are added to the feed to reduce the negative effects of metals deposited on the catalyst. The most widely known passivator is antimony which is used to reduce the dehydrogenation activity of nickel on the catalyst. This technology was developed by Phillips Petroleum during their early work on resid cracking. Antimony passivators are widely used by units with significant metals on catalyst. Chevron has patented the use of bismuth as a nontoxic alternate to antimony.

Various materials have also been used to passivate the effects of vanadium. These have generally been less effective than nickel passivators but recent developments show promise.

Vanadium Traps

Catalyst vendors have offered catalysts with vanadium traps for some years. These materials are intended to collect the vanadium in the feed and prevent it from reacting and destroying the zeolite in the catalyst. The traps are often incorporated into the catalyst but may also be sold as a separate additive.

References

[1]Butler, J.J., "Statistical Quality Control with CUSUM Charts," *Chemical Engineering,* August 8 (1983), pp. 73–77.

10
Troubleshooting

Troubleshooting FCC operations requires a firm grasp of the design and operating principles covered in the previous chapters of this book. Since problems with FCC operation usually result in significant decreases in refinery profits, FCC troubleshooting also requires a calm head and the ability to work under high levels of pressure to "get something done." When trouble strikes, FCC experts often appear—apparently from the woodwork as well as from dark places in the unit itself. Thus, the responsible engineer must not only cope with the actual problems at hand, but must also evaluate solutions and advice from various quarters.

Troubleshooting Basics

The basic steps in trouble shooting are: problem definition, analysis, problem solution, and implementation.

Problem Definition

The reported problem will often be related to some undesirable result; for example, a visible plume from the stack. The true problem, however, is the cause of the observed situation. In this case, it may be due to high catalyst losses or it may be due to high levels of sulfur oxides in the flue gas.

It is extremely important at this stage to evaluate not only the obvious problems but also the recent operation of the unit. In many cases, other changes will have occurred. These may be unrelated to the problem at hand, but they may also be either a second symptom of the root problem or may, in fact, be the cause.

Analysis

Once the actual problem is well-defined, possible causes are analyzed. In this step, ideas should be collected from as many sources as possible. Past experience is a valuable commodity at this stage. In many cases, similar problems may have occurred on this unit or on other units.

Once all possible causes have been collected, then each should be compared with the known facts. The expected consequences of each possible cause should be checked against the actual observations. This includes not only the current problem but also the other changes that have been observed on the unit. The most likely cause is the one that best explains all of the deviations from normal operations.

Problem Solution

Once the probable cause has been determined, solutions are evaluated. Again, past experience is a valuable tool. The solution may require operational changes and/or modifications to the unit. These should be discussed and evaluated by all concerned. In many cases, operations and maintenance will have valuable input as to possible alternate solutions that will minimize cost or down time.

Implementation

Once the solution has been established, it must be implemented. Where only operational changes are required, this can be done on line. If modifications to the unit are required, it may be necessary to shut-down operations. In this case, it will be necessary to schedule the shut-down so as to minimize the impact on the refinery. As in the previous step, operations and maintenance can offer valuable insight as to the best way to proceed.

Problem Areas

Many problems in the FCC fall under one or more of the following categories:

high catalyst losses
yield loss
poor catalyst circulation
afterburning
erosion of reactor/regenerator internals
coking of the reactor and transfer line
main fractionator fouling

Each of these is covered in some detail in earlier chapters but are summarized here for ease of reference.

High Catalyst Losses

High catalyst losses can be the result of cyclone damage, increased catalyst attrition, and/or changes in unit operation. Increased losses usually show up as either a need to increase catalyst make-up rates to maintain bed levels or as a decrease in the required catalyst withdrawal rate.

Increased losses from the regenerator may appear as an increase in stack opacity. If, however, the unit is equipped with third-stage catalyst separators and/or electrostatic precipitators, the increased losses may be collected and not appear in the stack. In these units, the higher losses will show up as a need to empty the collection hoppers more frequently.

Increased losses from the reactor will result in an increase in the catalyst content of the liquid drawn from the bottom of the main fractionator. If the fractionator bottoms product is not processed to remove this catalyst, these losses will also show up as an increase in the basic sediments and water (BS&W) of this product stream. If, however, the bottoms product is processed through filters or in a slurry settler, the increased losses may not result in an increase in the catalyst content of the product. Instead, it may be noticed as a need to backwash the filters more frequently.

If the increase in losses is associated with an increase in feed rate, a likely cause is increased velocities in the unit and thus, increased entrainment to the cyclones. High losses from the regenerator may be due to an increase in flue gas rate due to changes in coke yield or other operating parameters.

The level of fines (<40) in the catalyst inventory should be checked. If this level shows a decrease, the increased losses are most likely due to high velocities or a problem in the cyclone system. If the fines level is increasing, the problem is due to an increase in catalyst attrition or an increase in the fines content of the fresh catalyst being fed to the unit.

High catalyst losses with a falling fines level may indicate the formation of small holes in the cyclones or the plenum. Over time, these holes will be enlarged through erosion and the catalyst losses will increase gradually over time. If, however, the catalyst losses remain constant over time, the most likely cause is a plugged dipleg or a damaged trickle valve. If the losses are to the flue gas stack and the stack appears to be puffing, the most likely cause is dipleg flooding. This may be due to an obstruction in the dipleg or due to high dipleg levels caused by high velocities.

Increased losses with an increase in the fines content of the circulating catalyst may indicate increased attrition. This may be due to increased gas jet velocities in the unit. Increased gas jet velocities result from increases in the gas rates or from damage to gas distributors.

Increased attrition may also be due to a change in the attrition resistance of the fresh catalyst. This may indicate a manufacturing problem at the catalyst plant or some other change in the manufacturing processes.

An increase in the fines content of the fresh catalyst can also result in increased losses and an increase in the fines content of the equilibrium catalyst.

Losses from the reactor to the main fractionator may indicate a problem with coking in the cyclones or diplegs. This is a common problem in reactors with two-stage cyclones. In some cases, coke which forms in the cyclone or plenum will fall into the dipleg causing an obstruction. When this occurs, the catalyst losses appear suddenly over a very short period of time.

Poor Catalyst Circulation

Poor catalyst circulation usually manifests itself as a loss of slide valve differential or an inability to control the reactor temperature or the stripper bed level. Most catalyst circulation problems are the result of problems in the standpipes or at the standpipe entrance.

Since catalyst properties can have a major effect on standpipe operation, the catalyst G factor should be checked to confirm that it has not changed. A ten percent decrease in the G factor is often sufficient to produce noticeable circulation problems. A twenty percent decrease can result in severe circulation upsets on some units. If the G factor has changed due to a decrease in catalyst fines, this may indicate an emerging catalyst loss problem which must be addressed.

Large changes in catalyst circulation due to changes in operation can also result in circulation difficulties. This is especially true on units with long vertical standpipes. If a change of this magnitude has occurred, it may be necessary to retune the standpipe aeration.

If changes in the catalyst properties or unit operation cannot be identified as the probable cause of circulation difficulties, then a single-gauge pressure survey should be performed to locate the problems. Gamma scans and/or radiotracer studies may also be of value.

Yield Loss

Yield loss may be due to changes in feedstock, catalyst problems, or damage to unit internals.

Feed properties should be checked on a regular basis. This is especially true in refineries which switch feeds on a frequent basis. An increase in the specific gravity of the feed will generally indicate a higher aromatic content. This will result in poorer yields. Since basic nitrogen is a temporary catalyst poison, an increase in feed nitrogen will also generally result in a yield loss. Increased carbon residue will result in an increase in delta coke and this will cause lower conversions and reduced liquid yields.

These changes will be noticeable soon after the feed is introduced into the unit. Increased feed metals will also cause yield deterioration. This change will, however, happen gradually as the metals build up on the equilibrium catalyst.

Catalyst related yield losses can generally be identified by a change in one or more of the catalyst properties. Low catalyst activity can cause a loss of conversion and lower liquid yields. Activity loss may be due to hydrothermal or thermal deactivation or may indicate an increase in the level of vanadium on the catalyst. The actual cause can often be pinpointed by analyzing the catalyst's physical and chemical properties.

If the vanadium level on the catalyst has not increased significantly, a decrease in surface area, coupled with little or no change in the pore volume, indicates increased hydrothermal deactivation. This indicates an increase in the steam partial pressure in the regenerator which may be due to a leaking steam purge or an increase in the hydrogen content of the coke leaving the stripper. The latter of these would indicate a stripper problem.

A decrease in surface area coupled with a decrease in pore volume indicates thermal deactivation. This may indicate the formation of hot spots in the regenerator bed or an increase in afterburning. Either of these would point to damage to the regenerator air distributor.

An increase in the coke and/or gas factor indicates a loss of selectivity to desirable products. In the absence of a significant change in the catalyst activity this would indicate an increase in the level of nickel or some other dehydrogenation catalyst. In addition to nickel, copper and iron can increase dehydrogenation.

Since catalyst properties tend to show considerable random variation, statistical monitoring methods such as CUSUM analysis are valuable as aids to early detection.

An increase in gas or delta coke that cannot be explained by changes in feed or catalyst properties may indicate damage to the feed nozzles, riser termination device, or stripper.

Afterburning

Afterburning in the regenerator indicates that carbon monoxide is combining with oxygen in the flue gas. For units operating in complete CO combustion this indicates either inadequate air to the regenerator or poor air/catalyst distribution.

If the flue gas oxygen content has fallen recently, this indicates that either the air flow has decreased or the coke burn has increased, and there is inadequate air to the regenerator. If the air flow has been constant, then some other change has resulted in an increased coke yield. If the oxygen level has remained constant or increased, this is an indication of problems with air or catalyst distribution. The most likely cause is damage to the regenerator internals. If the pressure drop across the air distributor has changed, this is the likely problem area. A significant decrease in pressure drop may indicate cracks or other damage to the distributor. An increase in pressure drop may indicate partial plugging.

Erosion of Reactor/Regenerator Internals

Severe erosion of reactor/regenerator will usually be indicated by operational problems. These may include increased catalyst losses, poor stripping, poor feed atomization, localized afterburning, or incomplete catalyst regeneration. Erosion problems generally cannot be corrected without a unit shut-down. Their early identification, however, will permit better preparation and thus, minimize the loss of production required to make the necessary repairs.

Erosion or damage to the regenerator air distributor results in poor air distribution. This can lead to localized afterburning in the dilute phase or to hot cyclones. In addition, high local velocities may increase the overall catalyst attrition and catalyst losses from the regenerator. Poor air distribution may also produce a "salt and pepper" catalyst appearance.

Erosion of feed nozzle tips will result in poor feed atomization and distribution. This will be seen as an increase in dry gas yield and delta coke and possibly also result in increased transfer line coking. Erosion in the stripper will result in increasing hydrogen on coke and increased regenerator temperatures.

Erosion related problems generally increase gradually over time while mechanical failures produce step change effects. Internal damage from erosion or other causes can often be detected and located using radiotracer studies to track the distribution of catalyst and gas in various sections of the unit. Erosion problems are generally the result of high velocities or impingement of gas jets. These may be caused by increased feed rates or by other damage to the unit. Once erosion has been detected, it is critical that the cause be identified and corrected to prevent reoccurrence.

Reactor/Transfer Line Coking

Transfer line coking will be seen as an increase in the pressure drop between the reactor and the main fractionator. The most likely location for transfer line coking is at the transfer line flange and/or the entrance to the main fractionator. Other possibilities are pipe support attachments or any expansion joints in the transfer line. If the coke deposits are localized, the probable cause is excessive heat loss from that part of the transfer line.

Generalized coking over the length of the transfer line, however, indicates poor feed vaporization. This may be due to low reactor temperatures or to inadequate atomization in the feed nozzles. In these cases, there may also be severe coking in the reactor.

Some coke formation in the reactor is normal. These deposits will typically be found on tops of cyclones and other flat surfaces in the reactor. Severe, or widespread coking, however, indicates a problem with feed vaporization or with excessive reactor residence time. Heavy coking in the reactor dome (above the reactor cyclones) is caused by excessive residence time in this "dead zone."

Coke deposits are sometimes found on the outside of the reactor cyclone outlet tubes. These deposits usually form at a point 180° from the cyclone entrance. These deposits are fairly common and generally only produce problems if the coke breaks free and blocks the cyclone dipleg. This normally occurs as the result of a decrease in reactor temperature during a unit upset.

Main Fractionator Fouling

Main fractionator fouling usually occurs at either the top or the bottom of the tower. Fouling in the bottom of the tower is caused by the formation of coke. This is normally detected when pieces of coke block the bottoms pump suction strainers or bottoms circuit exchangers. Coke formed in the main fractionator can frequently be identified by the presence of a smooth flat surface on the coke particle. This surface indicates that the coke was formed on a metal surface in the tower. Coke of this type is usually hard and does not adhere to surfaces in the bottoms pumparound circuit. Soft, sticky coke found on the bottoms exchangers is normally formed in the bottoms circuit and not in the tower.

The hard coke formed in the tower indicates the presence of hot spots in the tower desuperheating section. These are most likely caused by poor liquid distribution, excessively high velocities in the transfer line or damage to the main fractionator internals.

Soft coke formed in the pumparound system may indicate excessively high temperatures and/or long residence times at high temperature. It may also be due to high conversion operations on aromatic feedstocks.

Damaged tower internals often produce symptoms in addition to coking. Often, the temperature profile in the bottom of the tower will have undergone a change at the time of damage. Increased temperatures above the desuperheating section coupled with decreased temperatures in the tower bottoms generally indicates damage that has reduced the vapor liquid contact in the bottom of the tower.

Fouling in the top of the main fractionator is due to an accumulation of salts, usually ammonium chloride. Typical symptoms are premature flooding, an inability to control the tower top temperature with reflux, poor fractionation, and, in some cases, the presence of salt deposits in pump suction strainers. Salt deposits are the result of low tower top temperatures and/or low reflux or top pumparound return temperatures. The most likely cause of these accumulations is operation at low overhead temperatures.

Turnaround Inspections

Turnaround inspections or T&Is are not normally considered a troubleshooting activity. They do, however, provide an excellent opportunity to inspect the condition of the FCC unit, and if they are done properly, can identify potential problem areas before they result in equipment failure.

This aspect of the T&I involves examining the unit not only for damage, but for indications of unusual wear or excessive stress. A trained and experienced eye is an essential element of this inspection as the indications of abnormal operation are often quite subtle and may be missed by a general inspection. The following list is intended to indicate most of the areas that should be given special attention. The list is quite extensive and it may not be necessary to inspect every item at each shut-down. A short list of special inspections can be prepared based on past experience in the refinery and the need to minimize turnaround time.

FCC Inspection Items

Riser
- inspect full length of riser for coke deposits
- pull and inspect feed nozzles for erosion and plugging
- inspect the wye steam ring for erosion and deformation
- check condition of refractory lining
- inspect refractory for erosion, spalling, and/or cracking (small cracks are normal)

Riser Separation Device
- inspect diplegs for coking and/or erosion
- inspect outlet tubes for coking, plugging, and/or erosion
- inspect separator body for erosion and for coke deposits in dead zones
- inspect refractory lining and anchors for damage (missing refractory, anchors pulled from walls)

Reactor Cyclones
- inspect diplegs for coking and/or erosion
- inspect trickle valves for erosion, damage, coking
- inspect refractory lining and anchors for damage or erosion
- inspect outlet ducts for erosion
- inspect cyclone body for coke deposits (coke often forms on the outside of the outlet duct)
- inspect hangers and supports for damage

Reactor
- inspect refractory lining for damage
- inspect pressure taps and thermal wells for erosion
- inspect dipleg bracing anchor points for erosion
- inspect dome steam ring for damage or coking
- inspect entire reactor vessel for coke deposits, remove coke where found.
- inspect vessel nozzles for corrosion and cracking.

Stripper
- inspect stripper vessel as a part of reactor inspection
- inspect baffles for damage and/or erosion

- inspect all steam rings for erosion, plugging or damage, inspect refractory lining on rings
- inspect debris guard on stripper exit for erosion or damage

Regenerator
- inspect refractory lining for damage
- inspect air distributors for erosion, plugging, and/or damage, inspect refractory lining on rings
- inspect pressure taps and thermal wells for damage
- inspect torch oil nozzles for damage and/or erosion

Catalyst Lift Lines
- inspect entire length of lift line for refractory damage and/or erosion
- inspect lift line exit for damage and/or erosion

Regenerator Cyclones
- inspect diplegs for erosion
- inspect trickle valves for erosion and/or damage
- inspect outlet tubes for erosion
- inspect cyclone interiors for erosion or damaged refractory
- inspect hangers for damage and/or erosion

Slide/Plug Valves
- disassemble slide valves and check for erosion on stems, guides, disks, throats, orifices and bodies
- remove plug valves, inspect for damage and for erosion on guide tube, stem tube guides, and stem tube

Expansion Joints
- check all expansion joints for damage and/or corrosion

Standpipes
- check refractory lining for erosion or damage
- check aeration taps and pressure taps for erosion
- check aeration taps and pressure taps for plugging and clear as necessary
- inspect J-bends for erosion, especially along the top of the lateral portion of the bend

Catalyst Coolers
- inspect tube bundles for damage, erosion, and deformation
- inspect fluidization air distributor for damage, plugging and/or erosion
- inspect cooler shell and refractory for damage and/or erosion
- inspect vent line for damage and/or erosion
- inspect cooler standpipes and lift lines

Flue Gas Line
- inspect flue gas line for erosion and/or refractory damage
- inspect quench nozzles for erosion, plugging, or damage
- inspect valves for corrosion or damage
- inspect any dead legs (e.g., expander by-pass lines) for corrosion

Third Stage Separator
- inspect separator vessel for damage and/or erosion
- inspect cyclones for damage, plugging, or erosion
- inspect underflow line for plugging or damage
- inspect critical flow orifices for catalyst deposits

Power Recovery Expander
- inspect turbine blades for excessive erosion
- inspect turbine blades for hot corrosion, and cracking
- inspect entire turbine for catalyst deposits (these usually form in areas of high velocity)

Reactor Overhead Vapor Line
- inspect refractory lining for damage
- inspect line for coke deposits

Main Fractionator
- inspect bottom section of main fractionator for coke deposits
- inspect top of tower for salt deposits

Unusual wear on any part of the reactor and regenerator should be noted and possible causes explored. This step is especially important during the first T&I on a new unit or on a unit that has undergone a major revamp. Unusual wear is often the first indication of an impending erosion problem that, if left uncorrected, could result in extensive damage.

A log of repairs should be maintained and reviewed after each T&I. If the same repairs are required at each shut-down, changes to the design of the equipment involved should be considered. This is especially true in areas subject to erosion. Repeated repairs to cyclones, air and steam distributors or riser termination devices indicate a problem with the design of these items. Correcting these design deficiencies will result in reduced turnaround schedules as well as improve the operational reliability of the FCC.

Glossary of FCC Terminology

As with most specialized fields of knowledge, catalytic cracking has developed an internal vocabulary which is used to amaze and confound the uninitiated. The following list is by no means all-inclusive, but it does contain definitions for most of the common mysteries as well as for common everyday engineering terms that have special meaning when applied to FCCUs.

ABD (Average Bulk Density). Density of a catalyst sample that has been allowed to settle undisturbed.

Activity. The conversion produced by a catalyst when tested on a specified feed at specified conditions. This is normally done in a bench scale test unit.

Afterburn. The increase in temperature between the regenerator dense bed and the dilute phase, cyclones or plenum chamber. This temperature increase is caused by the combustion of carbon monoxide to carbon dioxide.

Air Distributor. The device used to distribute combustion air across the regenerator bed. Air distributors may be grids, rings or pipe grids.

Aniline Point. The temperature at which a hydrocarbon sample becomes completely miscible in an equal volume of aniline.

Higher aniline points indicate a less aromatic sample. The aniline point is used in some FCC feed characterization methods.

Antimony. Antimony is used to passivate nickel deposits on catalyst and thus, decrease their dehydrogenation activity.

API Gravity. An expanded density scale based on specific gravity. API Gravity is expressed in °API and is calculated by the following formula:

$$°API = \frac{141.5}{\text{Specific Gravity}} - 131.5$$

APS (Average Particle Size). The weight average particle size of a catalyst sample. In many cases, however, the reported average particle size is in fact the median particle diameter.

Attrition Index. A measure of the resistance to the tendency of FCC catalyst to break up in the unit. There are several tests in use and comparisons can only be made between catalyst tested by the same procedure.

Backmixing. Localized flow of catalyst or vapors that is in the opposite direction to the overall flow. In the riser, catalyst backmixing contributes to higher yields of dry gas and increased delta coke.

Basic Nitrogen. Nitrogen that will react with the acid sites on the catalyst. Basic nitrogen is a temporary catalyst poison. The deactivating effects of basic nitrogen are reversed in the regenerator.

Beta Scission. The primary catalytic cracking reaction. Beta scission occurs when the carbon-carbon bond located one carbon away or beta to the charge site breaks in a carbenium atom. The products of beta scission are an olefin and a new carbenium ion.

Binder. Material used to hold together or bind the various constituents in FCC catalyst. The most common binder is sodium silicate.

Carbon on Regenerated Catalyst (C_{rc}). The carbon remaining on FCC catalyst after regeneration expressed as weight percent.

Carbon Snowball or Carbon Runaway. Build-up of carbon on the circulating catalyst due to inadequate regeneration air. Carbon snowballs often lead to high temperature excursions in the regenerator if the air rate is increased to burn the excess carbon from the catalyst.

C/O (Cat to Oil Ratio). The catalyst circulation rate expressed in mass/time divided by the feed rate expressed in the same units. Cat to oil ratios can be expressed in terms of fresh feed (FF) or in terms of total feed (TF).

Catalyst Circulation. The mass flow of catalyst over time. Usually expressed as tons/minute (T/M). Catalyst circulation is not measured directly but is calculated from the regenerator heat balance.

Catalyst Distributor. Any of various devices used to distribute catalyst across the regenerator cross section. Catalyst distributors are used to insure even distribution of the fuel (coke) burned in the distributor.

Catalyst Coolers. External heat exchangers used to remove heat from the regenerator. Catalyst coolers are often found in residue cracking units. Older catalyst cooler designs used dilute phase catalyst flow through the cooler tubes and steam generation on the shell side. Modern designs use a fluidized bed of catalyst on the shell side and the steam generation occurs in the tubes.

Cat Naphtha. Gasoline range material produced in the FCC. Also called cat gasoline, debutanized gasoline or stabilized gasoline. May be broken down into light, medium and heavy cuts.

Closed Cyclones. Any of several systems where the reaction riser discharges directly into a cyclone set so that the product vapors do not enter the reactor vessel.

Coke. Hydrocarbons that enter the regenerator with the spent catalyst and are burned. The majority of this material is a solid deposit on the spent catalyst. This deposit consists of highly condensed multiring aromatic compounds. In addition to this solid material, coke also includes any hydrocarbon vapors that are not removed from the catalyst in the stripper. The heat produced by coke combustion provides the heat required to vaporize and crack the feed.

Coke Factor. The coke produced by a given catalyst in the microactivity test expressed as a ratio of the coke that would be produced by a reference catalyst at the same conversion. Since the reference catalyst used may be different at each testing laboratory, this factor can only be used to compare catalyst tested at the same location.

Conradson Carbon (Concarbon). The residue left behind following pyrolisis of an oil sample under specified testing conditions. This measurement is used to estimate the fraction of FCC feed that cannot be vaporized.

Conversion. A measure of the quantity of feed converted into lighter products. Conversion is calculated by subtracting the percent yield of material heavier than gasoline from 100. Standard conversion is based on a 430°F TBP cut point gasoline.

CO_2/CO. The volume ratio of carbon dioxide to carbon monoxide in the flue gas from the FCC regenerator. This ratio is a measure of the completeness of the combustion of carbon in the regenerator. In complete combustion regenerators the carbon monoxide in the flue gas approaches zero and thus, this ratio approaches infinity.

CO Boiler. A fired furnace used to burn carbon monoxide in FCC regenerator flue gas to carbon dioxide. Normally, additional or supplemental firing is used to insure that the CO is burned. The heat evolved from the combustion of carbon monoxide and any supplemental fuel is used along with sensible heat in the flue gas to generate steam.

Cyclone. A mechanical separation device used to remove catalyst particles from either the reactor product vapors or the regenerator flue gas. Cyclones work by forcing the gas stream into a spiral flow pattern. The catalyst particles in the gas stream then move to the wall where they are separated.

Decanted Oil. Strictly speaking this term applies to oil taken from the top or decanted from a slurry settler. In practice, very few FCCUs use slurry settlers today. By convention, the term decanted oil is now used to refer to the heaviest liquid product produced by the FCC. This product is drawn from the bottom of the main fractionator. Other names for this stream are slurry oil and fractionator bottoms.

Delta Coke. The coke burned in the regenerator expressed as weight percent of the catalyst circulation. The delta coke can be calculated by dividing the coke yield (weight percent of fresh feed) by the catalyst to oil ratio (fresh feed basis).

Dense Phase. A fluidization state where gas flows upward through a fluidized bed of solid particles.

Dilute Phase. A fluidization state where particles are dispersed in and transported by a flowing gas stream.

Dipleg. Pipe used to carry collected solids from the body of a cyclone back to a fluidized bed.

Dispersion Steam. Steam injected with the oil feed through the feed nozzles. Originally, the purpose of this steam was to disperse the oil droplets formed in the nozzle and prevent their coalescence. In most modern nozzle designs the steam also helps to atomize the oil.

Distillation. Laboratory test to measure the boiling range of hydrocarbons. Several distillation tests are used to monitor the quality of FCC feeds and products. ASTM distillations evaporate a fixed volume of sample at a controlled rate. Vapor temperatures are recorded at specified volumes of condensate recovery. ASTM D86 distillation is used for lighter materials. ASTM D1160 is a vacuum distillation procedure used for heavier oils.

The TBP or true boiling point distillation uses a fifteen stage distillation with a five to one reflux ratio to measure the boiling points of pure components in the hydrocarbon sample. The GC simdist simulates a TBP distillation using a gas chromatograph.

Dome Steam. Steam injected into the space above the reactor cyclones to prevent coking in this "dead" volume. Also called anti-coking steam.

Dry Gas. Hydrogen, methane, ethane and ethylene produced in the FCC reactor. Combined with inerts from the regenerator and unrecovered C_3s, the dry gas forms the tail gas from the FCC gas recovery unit. This stream is usually used as refinery fuel gas.

Dynamic Activity. A measure of the ability of a catalyst to produce conversion at a constant coke make. The dynamic activity is calculated from MAT data by dividing the second order conversion factor by the coke yield.

E-Cat. Short for equilibrium catalyst. The catalyst inventory in an operating FCCU.

Electrostatic Precipitator. Unit used to recover catalyst from regenerator flue gas by charging the catalyst particles and collecting the charged particles on oppositely charged collection plates.

Faujasite. Mineral name for the zeolites used in cracking catalyst. Faujasites occur in nature, but are extremely rare. The zeolites used in FCC catalyst are synthetic Y faujasites.

Filler. Inert material used in FCC catalyst to give the particles their physical properties. The most common filler is kaolin clay.

Fractionator Bottoms. Liquid product from the bottom of the main fractionator.

Gas Factor. The gas yield produced by a given catalyst in the microactivity test divided by the gas yield that would be produced by a reference catalyst at the same conversion. As with the coke factor, the gas factor can only be used to compare catalyst tested at the same location.

Gas Recovery Unit. Absorption, stripping and distillation columns used to recover light FCC products and to separate them by boiling range. Also called the gas concentration unit, vapor recovery unit or simply the gas plant.

Grid Seal. Flexible metal seal used to close the gap between air grids and the regenerator wall. Grid seals are also used where the grid is penetrated by either a lift line or standpipe. Grid seal failure is a common problem in older FCCs.

Heat of Cracking. Theoretically, the endothermic heat of reaction in catalytic cracking. In practice, the heat of cracking is calculated by heat balance and thus includes corrections for errors introduced by various assumptions made as a part of the heat balance procedure.

Hydrogen Transfer. Bimolecular reactions that transfer hydrogen between two hydrocarbon molecules occupying adjacent active sites on FCC catalyst. The best known hydrogen transfer reaction transfers hydrogen from a cycloparaffin to an olefin resulting in a paraffin and a cyclo-olefin.

K Factor. The K factor is an indication of the paraffin content of an oil stream. There are two K factors in general use. The UOP K (K_{uop}) and the Watson K (K_w). Both are defined by the following equation:

$$K = \frac{ABP^{1/3}}{\text{Specific Gravity}}$$

where ABP is the cubic average boiling point in degrees Rankine for the UOP K, and the mean average boiling point in degrees Rankine for the Watson K.

Lift Line. Term frequently applied to risers other than the reaction riser.

Light Cycle Oil. Diesel range material produced in the FCC reactor. Yielded as a stripped side cut from the main fractionator.

Main Fractionator. Complex distillation column used to cool the product gases from the FCC reactor and to condense and separate the heavier product streams. Products from the main fractionator include the fractionator bottoms product, light cycle oil, unstabilized naphtha and wet gas.

MAT. MicroActivity Test. A bench scale test of catalyst performance that is performed using a standard feedstock and test conditions. The term MAT is also frequently used to mean the catalyst activity as measured by the microactivity test.

Matrix. The nonzeolitic materials in FCC catalyst. The term matrix is also used by some to indicate the nonzeolitic active ingredients in catalyst.

Octane Number. A measure of gasoline quality, the octane number is an indication of a gasoline's tendency to preignite or "ping" under compression. The reported octane number is actually the percentage of isooctane in an isooctane/normal heptane blend that has the same tendency to ping as the gasoline tested. The tests are conducted in a special engine. The research octane number (RON) is based on a test conducted at low engine speed (600 rpm) while the motor octane number is conducted at high engine speed (900 rpm).

Overcracking. Cracking that produces shorter-than-desired hydrocarbon chains. Usually applied to mean the cracking of cat naphtha to lighter products such as LPG and gas.

Particle Density. The density of a solid particle. In catalytic cracking, this is the density of the catalyst particle. Since catalyst particles are porous, this density includes the effect of the pore volume.

Particle Size Distribution (PSD). The distribution of particle diameters. This is expressed as the weight percent of the particles that fall within a range. For FCC catalyst the ranges are usually:

0–20µ
20–40µ
40–80µ
80–100µ
100µ+

Plenum. Chamber used to collect the gas from multiple cyclones. Plenums are used in both the reactor and the regenerator and may be either internal or external.

Plug Valves. Valves used in place of catalyst slide valves in some units. Plug valves are located at the base of a riser or standpipe and open and close with a vertical motion.

Pore Diameter. The diameter of the pores in FCC catalyst. This property is not measured directly, but is instead calculated from the pore volume and the surface area.

Pore Volume. The volume in the pores of the catalyst. This can be measured by nitrogen adsorption, mercury adsorption or by water saturation.

Power Recovery Turbine (PRT). Expander turbine used to recover pressure energy from FCC flue gas. In most installations the recovered energy is used to supply all or part of the energy required to drive the air blower.

Pseudo Density. "Density" calculated for fluidized catalyst by dividing the pressure drop across the catalyst bed by the height of the bed. The pseudo density is actually a measure of the pressure drop of the gas phase as it flows through the catalyst.

Ramsbottom Carbon. Another measure of the fraction of feed that cannot be vaporized. The testing procedure is different from that used to measure Conradson carbon and thus, gives a different result.

Rare Earth. Chemically, any element in the Lanthanide series. Used for ion exchange on the zeolite in FCC catalyst. Rare earth exchange improves the stability of the zeolite by reducing the hydrothermal dealumination of the crystal.

Refractive Index. A measure of the change in direction of a beam of light passing through the interface between two substances. Used in the n-d-M method to estimate the carbon distribution in oils.

Refractory. Nonmetalic materials used in FCC vessels, risers and standpipes to provide erosion resistance and/or internal insulation. Depending on the type of refractory and the service it may be installed gunned, vibrocast or hand-packed into a suitable anchoring system.

Riser. Transfer line used to move catalyst upward in dilute phase. The reaction riser uses vaporized feed to transport catalyst into the reactor. In a modern FCC, all or most of the cracking reactions actually occur in this riser. Spent catalyst risers are used to transport spent catalyst into the regenerator using air. Used alone, the term riser generally means the reaction riser.

Selectivity. The yield of a product divided by the conversion. Catalysts are often evaluated in terms of their selectivity to certain products. This is usually done using yields and conversions from the microactivity test.

Sensitivity. The difference between the research octane number (RON) and motor octane number (MON) of a gasoline sample.

Skeletal Density. The density of the solid material in FCC catalyst. Skeletal density excludes the catalyst pore volume.

Slide Valve. Valve used to control the flow of catalyst or flue gas in an FCC. Catalyst slide valves are single-disk valves while most flue gas slide valves are double-disk valves.

Slip. The velocity difference between solid particles and gas in dilute phase transport. Also called the slip velocity.

Slip Factor. The ratio between the solids residence time and the gas residence time in dilute phase transport.

FCC Terminology **317**

Slurry. Originally, the catalyst-rich oil recycled to the riser from the slurry settler. Today, the term is used to describe any of various streams taken from the bottom of the main fractionator (see decanted oil).

Slurry Settler. Vessel used to remove catalyst from the fractionator bottoms stream through gravity settling. Overhead product from the settler was yielded as decanted oil product while the catalyst-rich bottoms stream—slurry oil—was returned to the riser. Slurry settlers are generally no longer needed and many have been retired.

Spent Catalyst. Catalyst that has participated in cracking reactions. The spent catalyst contains a layer of coke that must be removed in the regenerator. Spent catalyst is not truly spent, but contains considerable residual activity.

Standpipe. A means of moving fluidized catalyst downward against a pressure gradient. Standpipes are used to move catalyst from the reactor to the regenerator and from the regenerator to the base of the riser.

Stick-Slip Flow. Typical flow in a poorly fluidized standpipe. Stick-slip flow is so named because the catalyst flows downward in a series of stop-start jerks.

Third Stage Separator. External separator on FCC flue gas line. Third stage separators are so-called since the flue gas has already passed through two stage cyclones before leaving the regenerator. Also referred to as tertiary cyclones.

Third stage separators are used to protect power recovery turbines from catalyst particles and/or to reduce the solids emissions from the FCC.

Trickle Valve. Valve used to close the bottom of a cyclone dipleg and thus, prevent gas from flowing up the dipleg.

Unit Cell Size. The length of the smallest repeating unit in a zeolite crystal. The zeolite unit cell size is an indication of the active site density of the catalyst and can be used to estimate the intrinsic activity as well as the yield selectivity of the zeolite.

Unstabilized Naphtha. Liquid product from main fractionator overhead. Also called wild naphtha.

Weight Hourly Space Velocity. The mass flow of feed divided by the mass of catalyst in the reaction zone. For FCC risers, this is usually taken to be the catalyst to oil ratio divided by the catalyst residence time.

Wet Gas. Gas stream from the main fractionator overhead condenser. The wet gas stream contains the dry gas produced by the FCC as well as most of the C_3 and C_4 LPG.

Wet Gas Compressor. Compressor used to raise the wet gas stream to the pressure of the gas recovery unit.

Zeolites. The primary active ingredient in modern FCC catalysts. All catalyst today use synthetic Y-Faujasite. The zeolite is initially crystallized as a sodium Y or Na – Y. Ion exchange with ammonia and/or rare earth converts the Na – Y to either H – Y or Re – Y zeolites. H – Y zeolites can be converted to ultrastable Y or US – Y zeolites by heat treatment.

Index

A

Activity, 279, 307
  effect on yields, 298
  effective activity, 62, 282
  high alumina catalysts, 3, 4
  zeolite catalysts, 4, 263
Afterburn, 156, 307
  causes, 298, 299
  effect of air distribution, 144
  effect of carbon distribution, 143
Antimony passivator, 291, 308
Average boiling point (of feed), 313
  use of feed characterization, 58
Average bulk density (ABD), 211, 280, 307
  effect on fluidization, 275

B

Backmixing, 92, 94, 308
Beta scission, 308
  effect on yields, 48, 51
  in catalytic cracking, 47
  in thermal cracking, 51
Binder, 71, 308
  catalytic activity of, 267

C

Carbenium ions, 47–50
  role in catalytic cracking, 47, 48
Carbon on regenerated catalyst, 309
  effect on catalyst activity, 281
  effect of regenerator conditions, 150
Carbon residue, 53
  conradson, 310
  effect on yield, 5, 64
  ramsbottom, 315
Carbon snowball, 309
  causes, 156
  corrective action, 160, 161
Catalyst
  bed density, 203
  coke factor, 279, 280
  filler, 71, 264, 267
  gas factor, 279
  heat capacity, 76
  manufacture, 269–272
  monitoring, 278–288
  selection, 272–278
Catalyst additives, 289–291
Catalyst circulation
  calculation from heat balance, 76
  problems with, 214–217

Catalyst coolers
  in early units, 3,
  in resid cracking, 38, 40
Catalyst distributor, 145
Catalyst losses, 72, 186
  causes of high losses, 295
Catalyst matrix, 267–269
  active matrix components, 268
  surface area, 268, 280
Catalyst particle density, 188, 190, 202, 314
  calculation of, 204
Catalyst particle size distribution, 281, 314
Catalyst pore diameter, 315
Catalyst pore volume, 280
Catalyst skeletal density, 205, 212
Catalyst to oil
  effect on conversion, 62
CO/CO_2
  in partial burn regeneration, 157
CO boiler, 13, 165
Combustion promoter, 72, 289–290
Coke product, 64
  effect of catalyst residence time, 54
  hydrogen content, 74
  heat of combustion, 74
  yield calculation, 73–75
Coke deposits
  reactor, 113, 300
  transfer line, 223, 224, 255, 300
  fractionator, 226, 256, 300–301
Conversion, 59, 310
  effect of catalyst to oil ratio, 62
  second order, 62
  standard, 310
Cumulative sum of deviations (CUSUM), 286
  use in catalyst monitoring, 286–288
Cylcone, 183
  closed, 108
  collection efficiency, 190, 191
  erosion, 193
  in third stage separators, 170
  pressure drop, 192
  reactor, 12, 186
  regenerator, 13, 183–185

D

Distillation, 311
Decanted oil, 223, 311, 317
Delta coke, 311
Dense phase, 197
Dilute phase, 197
Dipleg, 147, 183
  back up, 192
  erosion, 193
  flooding, 296
  mass velocity, 187
  length, 193
Dispersion steam, 311
Dome steam, 113
  steam quality, 114
Dry gas, 312
Dynamic activity, 279–280, 312

E

Electrostatic precipitator, 20, 175–179
  use in catalytic cracking, 175
  safety issues, 179
Entertainment, 188
Equilibrium catalyst
  analysis, 282
  metals, 282, 283
  sampling, 285

F

Feed nozzles, 92, 93
  commercial types, 97–100
Feed stocks, 7, 52
  characterization, 57–59
  hydrocarbon types, 53–57
  types, 53

Flue gas
  composition, 163
  heat content, 76
  NO_x removal, 180
  particulate removal, 168, 176
  sulfur removal, 180

G

Gasoline
  effect of conversion on yield, 63
  sulfur content, 67, 69
Gas recovery unit, 236–249
  absorber/stripper as single tower, 244
  debutanizer, 247, 248
  primary absorber, 241, 242
  secondary absorber, 246
  stripper, 243–246
Grid seal, 135, 136

H

Heat balance, 72–77
  reactor, 76, 77
  regenerator, 75, 76
  heat of combustion, 75
Heat of cracking, 77
Hydrogen blistering, 238
  chemicals for control, 240
  role of cyanides, 238
Hydrogen transfer, 48, 49, 51
  catalyst effects, 265
  effect on gasoline octane, 265
  effect on yields, 265

J

J-bend, 26

K

K factor, 313

L

Lift line, 20, 143, 144
Light cycle oil, 59
  yield as a function of total cycle oils, 65
  sulfur content, 69

M

Main fractionator, 14–16, 225–236
  coking, 256, 257
  fouling, 258
  tray efficiencies, 235
Mass balance, 77–83
  adjustment to standard cuts, 81
  coke yield, 73
  validity for test run, 80
Mass spectroscopy, 58
Mercaptans, 251
  extraction from LPG, 251, 252
  sweetening in gasoline, 252, 253
Microactivity test, 273, 279
  use in catalyst evaluation, 273
  data from, 279, 280

N

Nitrogen oxides (NO_x), 180, 181
  control, 181
Nuclear magnetic resonance spectroscopy, 58

O

Octane number
  effect of operating conditions, 70, 71
  effect of feed quality, 71
  catalyst effects, 265
Orifice chamber, 164, 165, 167
  orifice plate design, 167
Overcracking, 314
Overflow well, 23, 24, 209

P

Partial combustion, 130, 163
   control of, 157
Plenum, 148, 315
Plug valve, 26, 27, 36, 315
Power recovery, 171–174
   early experience, 171
   effect on unit reliability, 174
Pseudo density, 203, 207, 214

R

Radiotracers, 217, 219
Rare earth, 265, 266, 270
Reactions
   catalytic cracking, 47, 48
   hydrogen transfer, 49, 51, 265
   primary, 50
   secondary, 49, 51
   thermal cracking, 50, 51, 52
Reaction mix sampling, 82, 83
Reactor vessel, 11, 12
   design, 112–115
Refractive index, 58, 59
Refractory lining, 114, 131, 138
Regenerator
   cyclones, 146–149
   design, 131
   two stage, 131–133
Regenerator air distributor, 134, 299, 307
   grids, 134, 135
   pipe grids, 135–139
   rings, 139, 140
   nozzle design, 140–142
Resid cracking, 5, 38–40
Residence time
   effect on coke, 64
   effect on conversion, 62
Riser, 4, 5, 10, 11
   design, 101
   external, 102
   internal, 102
   right angle turn, 103
Riser termination device (RTD), 104–112

S

Slide valve, 11, 13, 164, 165, 219, 220
Slip
   factor, 316
   velocity, 200, 206
Slurry, 223
Slurry settler, 317
Spent catalyst, 12
   residual activity, 42
Sponge absorber, 16, 246, 259
   effect of increased sponge oil rate, 247
   purpose, 246
Standpipe, 13, 205–217
   aeration, 211, 212
   density, 206, 207, 212
   design, 207–210
   hopper, 209
   overflow, 207
   underflow, 207
Sulfur
   in products, 66–70
   in flue gas, 163
   removal from flue gas, 180

T

Third stage separator, 13, 168–170
   use with power recovery, 171, 173
Transport disengaging height, 189
Treating, 18, 249
   gas, 249
   LPG, 250, 251
   naphtha, 252, 253
Trickle valve, 185

U

U-bend, 23, 24
Unit cell size
  effect of rare earth on, 266
  effect on octane, 265
  effect on yields, 265

V

Vanadium
  effect on catalyst, 282
  traps, 291

W

Water wash
  purpose, 238
  concurrent, 238
  countercurrent, 239
  parallel, 240

Weight hourly space velocity
  calculated from cat/oil, 62
  effect on conversion, 62
  effect on coke yield, 65
Wet gas, 238
Wet gas compressor, 16, 234, 236, 237

X

X-ray diffraction, 267

Z

Zeolite, 4, 71, 264–267
  hydrogen Y, 265, 270
  rare earth Y, 265, 270
ZSM-5, 72

It's easy to do the right thing.

CCC makes it simple, efficient, and cost-effective to comply with U.S. copyright law. Through our collective licensing systems, you have lawful access to more than 1.7 million titles from over 9,000 publishers. Whether it's photocopying, electronic use, or the emerging information technologies of tomorrow—CCC makes it easy.

Call 1-800-982-3887 ext. 700 to find out how CCC can help you to Copy Right![SM]

Copyright Clearance Center®
Creating Copyright Solutions
222 Rosewood Drive
Danvers, MA 01923

Copyright Clearance Center and the CCC logo are registered trademarks of Copyright Clearance Center, Inc. within the United States